Business Earth Stations for Telecommunications

WILEY SERIES IN TELECOMMUNICATIONS

Donald L. Schilling, Editor
City College of New York

Business Earth Stations for Telecommunications

Walter L. Morgan
Consultant
Communications Center

and

Denis Rouffet

WILEY

A Wiley-Interscience Publication
JOHN WILEY & SONS
New York · Chichester · Brisbane · Toronto · Singapore

Library of Congress Cataloging in Publication Data:

Morgan, Walter L.
  Business earth stations for telecommunications / Walter L. Morgan
  and Denis Rouffet.

        p.   cm.—(Wiley series in telecommunications)
  "A Wiley-Interscience publication."
  ISBN 0-471-63556-1
  1. Business—Communication systems.   2. Earth stations (Satellite
telecommunication)   I. Rouffet, Denis,   II. Title.   III. Series.
HF5541.T4M67   1988
658′.05—dc 19   87–29055
                                                           CIP

Printed in the United States of America

10 9 8 7 6 5 4 3 2 1

PREFACE

This book describes the practical applications of these small earth stations for business telecommunications. It examines several typical uses in sufficient depth to allow readers to configure their own networks to fit their particular needs. Alternatives are explored because due to the flexibility of this technology no single design can satisfy all potential users.

AUDIENCE

Microterminals offer corporate and government telecommunications managers a special opportunity to regain control of telecommunications costs and network operations while providing a flexible array of services. This book is intended to be a practical guide to the evolving requirements for microterminals. It also covers the needs of satellite system operators and manufacturers who want to design satellites that can capture the microterminal traffic. Government officials wanting to understand what they are being asked to regulate will find information on the various present and future prospects for this industry. Private international microterminals are appearing and may be regulated in a manner quite different than huge public earth station networks. This book provides insight into the planning needs of manufacturers, sellers, and installers of this type of equipment for the next decade.

As with any new technology, there is need for a multitude of peripheral equipment (codecs, protocol converters, switching, control centers, etc.) which is of interest to manufacturers looking for expansion opportunities.

Microterminals may not be an optimum solution for every corporate telecommunications manager. We hope this book will be a valuable aid to readers in making such decisions.

Another use of this book may be as a supplementary text in both business and engineering schools.

In addition to the pure VSAT data applications, uses for voice and facsimile are explored. The microterminals can also be used to receive conventional analog television for employee training and communications.

The text considers the use of these terminals at the data rates of 1.544 Mb/s (T1 in North America) and the CEPT counterpart at 2.048 Mb/s in Europe. At these rates the applications expand to include business teleconferencing and switched (at the hub) telephony or data transmission. The text is designed to recognize future applications for these terminals. Cost and facility projections are included.

Special requirements of hub earth stations and satellites change as the uses of the microterminal networks change. This book examines how the shared hub station operator improves the economics of the smaller networks by allowing the cost sharing of the most expensive parts of a network.

This book originally began as a Communications Center report on the application of microterminals in the United States and their expansion to Western Europe. The report was prepared for the French space agency, the Centre National d'Etudes Spatiales (CNES), in Paris, under the direction of Arianespace, Inc., Washington, D.C. At the completion of the study, it was suggested that a shortened version would be a useful contribution.

We acknowledge the assistance of Kim Kaiser (who built the worlds's first working microterminal—see Figure 3.3), especially in the sections dealing with spread spectrum and protocols. The Communication Center's clients at GTE Spacenet (Jerry Waylan), M/A-Com (Len Golding), Federal Express (Johnny Wong), and Arianespace, Inc. (Douglas Heydon and Diana Josephson) provided valuable reviews. The Communications Center's Clarksburg Maryland staff participated at all levels; special thanks are due to Peggy Petronchak and Carol Streeter.

We dedicate this book to our families.

WALTER L. MORGAN
DENIS ROUFFET

Clarksburg, Maryland
Paris, France
December 1987

CONTENTS

CHAPTER 15 MICROTERMINAL INSURANCE ASPECTS 217

LIST OF TABLES

LIST OF FIGURES

1

GENERAL DESCRIPTION

1.1 A BRIEF HISTORY OF SATELLITES

As shown in Table 1.1, the early satellites used low orbits that passed quickly over an earth station. The stations needed large antennas with high-speed tracking capabilities, which greatly increased their cost. The low-altitude satellites had limited coverage because of their height. Communications between two stations could only be handled for limited periods of time when both stations (e.g., United Kingdom and Maine) could see the same satellite. If Maine and Chicago wanted to communicate, another satellite orbit would be needed.

Since the orbits were relatively short (e.g., 100–200 minutes), there were long periods when the satellite could not be seen by the earth station. To provide continuous service would require many, many satellites. Multiple antennas were needed at each earth station if multipoint communications were needed, because different satellites were needed due to the mutual visibility requirement.

In 1963, the first geosynchronous satellite (*Syncom II*) was launched. It viewed the entire surface of the earth that it could see (approximately one-third of the total), but had an inclined orbit that had to be tracked from the earth stations. While this made the stations considerably lower in cost than for low-orbit satellites (and only one antenna would be needed as opposed to many), these antennas still needed the capability of tracking the moving satellite. *Syncom II* had an orbit that was synchronized with the rotation of the earth and therefore had a period of a sidereal day (about 24 hours). It was in an inclined orbit, which meant that the satellite moved north and south of the equator during the day.

1

TABLE 1.1   History of Satellites

First Launch	Type of Orbit	System Example	Coverage Area	Earth Station Antenna Size[a]	Earth Station Antenna Tracking	Rel. Cost[b]
1960	Low altitude	Echo	Brief and limited	Large	High speed	Costly
1962	Medium altitude	Telstar and Relay	Brief and limited	Large	Yes	Costly
1963	Geosynchronous	Syncom II	Full	Large	Yes	Loss costly
1964	Geostationary	Syncom III	Full	Large	Limited	Less costly
1965	Geostationary	Early Bird Intelsat I	Full	Large	Limited	Less costly
1972	Geostationary	Anik AI domestic	Canada	Medium	Limited	Much less
1973	Geostationary	Westar I	US	Small	None	Low
1974	Geostationary	Satcom I	US	Small	None	Low
1976	Geostationary	CTS-1[c]	U.S./Canada	Smaller	None	Low
1980	Geostationary	SBS-1[c]	U.S.	Small	Very limited	Varies
1985	Geostationary	Gstar I[c]	U.S.	Smaller	None	Low
1985	Geostationary	Satcom K[c]	U.S.	Smaller	None	Low

[a]Sizes (diameter): large, 20–30 m; medium, 15 m; small, 5–10 m; smaller, 1.2–3 m.
[b]Relative cost (1987 dollars): costly, in excess of $10 million; less costly, $2–10 million; much less, under $2 million; low, under $25,000.
[c]Operates at 14/12 GHz (K_u-band).

A year later, *Syncom III* was launched into a geostationary orbit. In this case, the satellite must have an inclination angle near 0° and an orbital period that again matches the earth's rotation. This combination makes the satellite appear to hover over a particular spot on the earth's equator. These satellites are 22,300 statute miles above the equator. Since the satellites are not exactly motionless, and the large earth station antennas had narrow antenna beams, a limited tracking capability was necessary to keep the antenna pointed exactly at the satellite. The large antennas were needed because the satellite had a low-power transmitter.

The first commercial application appeared in 1965 with the launch of *Early Bird (Intelsat I)* for international service. The standard antennas for this service were 30 m (nearly 100 ft) in diameter and used cryogenically cooled receivers. Only large common carriers could afford to build and operate these stations.

The early 1970s saw the emergence of domestic satellite services for Canada and the United States. These satellites had medium-powered transmitters and concentrated that power on only part of the earth (e.g., Canada or the United States) and were more sensitive. The satellite motion was tightly constrained by the use of on-board thrusters and fuel. This permitted the use of small and medium-sized earth stations with either very limited or no tracking capability.

All of these satellites operated at C-band (uplink at 6 GHz and downlink at 4 GHz). A gigahertz is 1000 MHz; 4 GHz is approximately 40 times the frequency of a standard FM radio station at 100 MHz.

In the late 1970s, NASA and Canada launched CTS-1, a high- and medium-powered satellite to operate with small earth stations. It proved the feasibility of service at 12 and 14 GHz (the ''K_u-band'' into small and transportable terminals).

In 1980, the first commercial K_u-band satellite appeared in the form of *SBS-1*. The original *SBS-1* antennas were expensive, but eventually, smaller and lower-cost antennas were deployed with very limited tracking.

The mid-1980s saw higher-powered satellites (e.g., *Gstar* and the *Satcom-K* series), which permitted still narrower antenna beams (half of the United States). This combination of more radio-frequency (RF) power in the satellite, and a more concentrated beam permitted the use of still smaller earth station antennas which cost much less. These small antennas had sufficiently broad beams that no tracking was necessary.

The K_u-band services also had the added advantage of the absence of significant terrestrial interference. In contrast, the C-band stations are in the same frequency range as the point-to-point terrestrial microwave services of many telephone and video common carriers.

1.2   A BRIEF HISTORY OF SMALL EARTH STATIONS FOR BUSINESS

Table 1.2 shows a similar chronology for earth stations. The earliest satellites were so weak that they required giant 100-ft antennas of the original *Intelsat* Standard-A Class. The domestic satellites used substantially smaller antennas. In the mid-1970s, experiments were conducted with Dow Jones (the *Wall Street Journal*) which showed that small antennas (10 m) could be used for commercial purposes. Prior to this time, almost all of the service was either bulk telephony or television services.

At the same time, experiments were being conducted on the hospital ship SS *Hope*, which was anchored in Brazil. After the medical experiments were completed, the 2.4-m antenna and equipment were removed from the ship and taken for a tour of Brazil, including up the Amazon River. This demonstration by Joachim Kaiser of COMSAT Laboratories showed that telecommunications could be very important for a large and growing nation such as Brazil. These experiments in 1974 led to the eventual Brazilian national satellite system and the growth of an earth station business within Brazil.

Two years later, COMSAT Laboratories designed and deployed a rapidly transportable station for rescue and emergency operations using a 1.2-m an-

TABLE 1.2   History of Earth Stations

Year	Class of Customer	Antenna Diameter		Service Capability (Voice Circuits)
		Meters	Feet	
1962	AT&T Telstar	30	100	Tens
1965	Intelsat	30	100	Hundreds–thousands
1973	Westar	15	50	Thousands
1974	Cable TV	5–10	16–32	TVRO[a]
1974	*Wall Street Journal*	10	32	Newspaper
1974	SS *Hope*	2.4	8	1 (to a ship)
1976	Rescue	1.2	4	1 voice circuit
1978	Business (ASC)	4.5	15	56-kb/s SCPC
1979	Business	10	32	1.544-Mb/s SCPC
1980	Business (SBS)	5.5	18	40-Mb/s TDMA
1981	One-way data	0.6	2	300–9600-b/s SSMA
1984	Two-way data	1.2 × 0.6	4 × 2	1.2–19.2-kb/s SSMA
1985	Two-way data	1.2	4	56 kb/s

[a]TVRO = Television receive only. See Glossary for other definitions

tenna. It used 20- or 200-W power amplifiers in the experimental *CTS-1* satellite that each had a spot beam that could be moved by remote control to link cities in Canada or the United States.

The American Satellite Company was busy applying what was then referred to as small earth station technology to business applications at both 56-kb/s and 1.544-Mb/s services using 4.5- and 10-m-diameter antennas and unmanned stations that operated with the *Westar* satellites.

In 1981, Equatorial Communications Company started selling the first of tens of thousands of one-way data terminals using 2-ft receive-only antennas. Three years later, they deployed a larger two-way data terminal. Both of these used spread-spectrum multiple access at 4 and 6 GHz.

It was not until the mid-1980s that higher speed (56 kb/s) was deployed through still smaller antennas at 12 and 14 GHz using other techniques.

The beginning of small business earth stations may be traced directly to the Request for Proposal issued by Federal Express for 25,000–50,000 microterminals for use in their corporate and planned facsimile networks.

1.3 SATELLITE CATEGORIES

The three categories of interest for business earth stations are officially called fixed satellite service (FSS), the broadcasting satellite service (BSS), and the mobile satellite service (MSS).

The terms "fixed satellite" and "mobile satellite" have nothing to do with the motion of the satellite. The nomenclature describes how the earth station antennas are mounted on the ground. Fixed satellite service stations are generally intended to be mounted to an immovable object on the earth (a building or the ground). Mobile stations are placed on ships, airplanes, automobiles, and so on. Transportable earth stations (which are intended to be moved from one fixed location to another, but not to communicate enroute) are licensed as fixed satellite service stations. Broadcasting satellites are intended for reception by the general public. This is commonly interpreted to mean directly to the home via an on-premises receiving antenna.

This book deals with earth stations that are mounted to the earth (at least during the time of their use). Since these terminals are so small, there is nothing to prevent them from being removed, placed in a truck, and moved to another location where they may be reinstalled.

Business services to date have been restricted to the fixed satellite services. Construction of facilities in the other bands may be permitted under certain conditions. The broadcasting satellite service band at 12.2–12.7 GHz, for example, may be used on a secondary basis for nonvideo services.

1.4 SPECIAL TERMINOLOGY

1.4.1 Business Telecommunications

Digital Speeds. Table 1.3 shows the various digital data speeds commonly used in North America and Europe. The most common speeds for business earth stations are the 300–1200 b/s used by many personal computers and word processors, 9600 b/s for more sophisticated computers and terminals, and the range 56–64 kb/s (combined in this book) for medium-speed data (including combinations of many lower-speed terminals). The highest speeds considered in this book are the T1 (1.544 million b/s) or the European equivalent at 2.048 million b/s. As Table 1.3 shows, there are other speeds.

Analog Services. The primary analog service in business telecommunications is voice. Through the use of analog-to-digital encoders, the voice may be sampled 8000 times per second. Each sample may be converted from the analog domain (where voltage varies as a function of the loudness of the speaker's voice) into a digitized value using 7 bits. In the end, 56 kb/s was originally required. Through the use of more sophisticated encoding techniques, modern technology now permits placement of two, three, or even more voice channels in the 56 kb/s originally reserved for just one. Once the voice has been encoded, it appears like any other digital data and may be mixed with digital bit streams from computers, point-of-sale devices, word processors, and so on. In a similar manner, other primarily analog domain sources (such as business to business television and the older facsimile systems) are rapidly being converted to digital for improved transmission and reduced circuit requirements (which therefore lowers their transmission cost).

TABLE 1.3 Common Digital Data Speeds

Rate[a]	North American Designation[b]	Primary Use
Under 300 b/s	TTY	Teletype
300 b/s	—	Low-speed data to microcomputer
1200–2400 b/s	—	Low-speed data to microcomputer
4800–9600 b/s	—	Medium-speed data to microcomputer
56–64 kb/s[c]	T0 (also Europe)	Medium-speed data, voice and compressed video
144 kb/s	—	Basic ISDN (2B + D)
1.544 Mb/s	T1	High-speed data or multiple voice channels
2.048 Mb/s	E1 (Europe and non-U.S.)	High-speed data or multiple voice channels
3.152 Mb/s	T1C	Two T1's
6.312 Mb/s	T2	High-speed data or multiple T1's
44.736 Mb/s	T3	Bulk digital telephony

[a]$k = kilo = \times 1000; M = mega = \times 1,000,000.$
[b]In place of T, other prefixes (e.g., D, DS, FT, etc.) may be employed for specific transmission forms.
[c]Lower-speed (e.g., 16 or 32 kb/s) bit rates may be used for voice or the T0 or T1 may be subdivided into voice and/or slower-speed data channels.

Processing Systems. It is common to talk about data processing (which deals primarily with numbers) and word processing, but there is also image processing in the form of both facsimile and television. These systems can manipulate the digital information (whether it be derived from a data or analog source) through the use of switching, enhancements, regeneration, and so on. Unlike the analog telephone systems of the past, these telecommunications links use integrated digital circuits and the power of microprocessors. Transmission errors may be found and corrected to provide outstanding performance unmatched by analog facilities.

Interconnections. The world of telecommunications can be simply divided into two forms: public switched networks and private networks. In the case of the public switched network, access into the network is permitted to any subscriber. This is typical of the dial-up telephone systems. Egress from the network is also permitted to any subscriber. This permits the interconnection of large numbers of subscribers.

Private networks are intended for direct connection between parts of the same company or between various companies. In general, the private and public networks cannot access one another. This is done primarily for regulatory purposes.

Packet switching is a transmission method which involves the digital telecommunications from many users being bundled together to be handled in more efficient manners. These networks also have the advantage of being highly reliable through the use of many alternative paths so that if one circuit gets too busy or goes out of service, an alternative path can be selected by the switch on an individual packet basis. The user is not aware of how the telecommunications reaches the destination.

1.4.2 Satellite Communications Terminology

Once the signals within a business have been collected, prioritized, switched, combined, and identified, they are ready for impression on a radio-frequency carrier. This process is called modulation.

A satellite transponder (the repeater in the sky) has sufficient capacity in a single transponder to handle many simultaneous messages from earth stations. The process by which this entry into the satellite system is permitted is called multiple access. Modulation and multiple access are different processes.

Many business systems using small earth station antennas (defined as less than 2.8 m for the purpose of this book) are too weak to communicate with one another except through a large master station, commonly referred to as a hub. If the individual small or microterminals are thought of as being arranged in a circle around the hub, the communications lines between the small antenna and the hub look like a star. From this perception comes the term "star networks." The hub provides the power boosting (and often modulation changing) needed to permit the small stations to "talk" to one another. The hub also provides the important function of managing the network, monitoring its performance, and keeping track of who should be billed how much at the end of the month for use of the services.

In many respects, the hub resembles a highly intelligent telephone switching office. In addition to the radio-frequency equipment, the hubs (and many times the microterminals) require the use of computing facilities, switching capabilities, business management systems, troubleshooting capabilities, and so on.

1.4.3 Microterminal or "V-SAT"

The term "V-SAT" is a trademark that was used by Telecom General for their series of earth stations. In this book we use the more generic term "microterminal."

1.5 BEYOND THE UNITED STATES

Microterminals are being permitted in most major single-country domestic satellite systems, including Canada, Mexico, and Australia. Many of the U.S. applications will be used in other countries. Less-developed countries, in particular, may find the use of these terminals particularly attractive as a means of quickly upgrading existing telephone systems and extending capacity to corners of the nation that were previously served by high-frequency radio or other methods. Microterminals permit rapid expansion of a network to handle temporary traffic surges or growth.

1.6 THE ROLE OF MICROTERMINALS IN A FIBER OPTICS ERA

Fiber optics and microterminals serve opposite ends of the telecommunications spectrum. Fiber optics is exceedingly efficient (both operationally and economically) when it comes to handling massive numbers of telephone circuits of very high speed data rates (in the range of hundreds of millions of bits per second). Therefore, fiber optics, along with teleports and major earth stations, is used to tie the lines between minor facilities (consisting of thousands of people).

Microterminal networks operate best with a few voice channels or moderate digital speeds. These networks are used to tie many small, outlying locations (e.g., department stores, fast-food outlets, service stations, etc.) into the headquarters.

Satellites may also be used to distribute business teleconferencing video efficiently on a point-to-multipoint basis by exploiting the inherent broadcast nature of the satellite. In some networks, the same satellite (and generally the same polarization to simplify the earth stations further) is used for both microterminal data and teleconferencing (usually, the two services are different transponders).

There are opportunities for both types of services. In the business environment there are many large companies that use fiber optics and there are tens of thousands of small business operations (including the branch offices of large corporations) that may be more efficiently served through the use of microterminals.

2

WHAT IS A MICROTERMINAL?

This chapter is an overview of the microterminal: what it is, how it is operated, and so on. Many details will be deferred to later chapters.

2.1 TYPICAL MICROTERMINAL AT 14/12 GHz

A typical microterminal for use in the 14/12-GHz frequency band (K_u-band) for microterminal or VSAT* network applications consists of two major parts. The first is the outdoor unit containing the antenna reflector, feed system, the RF electronics package, and the antenna mount (see Figure 2.1). The other part is the indoor unit containing the signal processing equipment and customer interfaces.

2.1.1 The Outdoor Unit

The reflector can be made out of metal (aluminum) or metallized plastics. Mesh antennas, commonly used for low-cost C-band antennas, are rarely found in professional installations, especially at K_u-band. The reflector is normally a parabola (or a part of a parabolic surface with a feed system that is offset with respect to that surface). In the offset case the feed is arranged in such a way that it is not in the beam of the antenna. Advantages of this arrangement over a prime-focus front feed design is that the offset feed does not block the beam. Blockage reduces the available antenna gain and produces undesirable sidelobes in the antenna pattern.

*Very small aperture terminal.

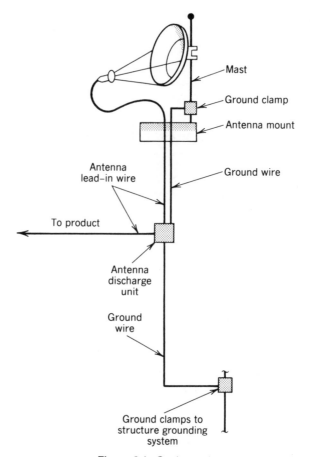

Figure 2.1. Outdoor unit.

The feed is rigidly attached to the reflector by suitable means, for example, three spars attached to the reflector edges or, in many cases, by a cantilevered beam attached to the mount and/or the reflector. The feed must form a rigid unit with the antenna surface (see Figure 2.2). The RF electronics unit can be connected to the feed system in several ways. Some designs integrate the feed and the RF electronics into a weatherproof enclosure; others connect the feed to the electronics box by waveguide and/or coaxial cables. To avoid serious signal losses at 12 and 14 GHz, coaxial cable or waveguide lengths must be held to very short runs, typically less than 30 cm or 1 ft. For this reason, the frequency of the inbound and outbound signals are converted to, or from, a lower (intermediate)-frequency (IF) as close to the feed as possible. Typically, the IF is 950–1450 MHz.

The outdoor units are conditioned to withstand the wide variations in wind,

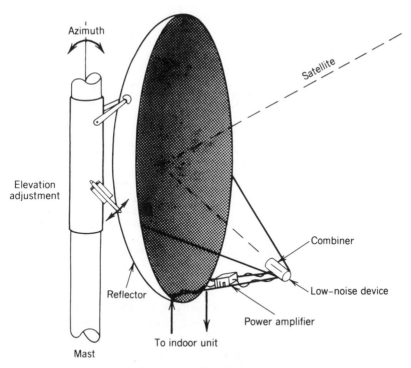

Figure 2.2. Offset-fed antenna.

temperature, and humidity conditions. Heat generated by the electronics is normally passively dissipated by means of judiciously placed radiating fins. Typical data terminals use 1- to 2-W solid-state power amplifiers. In the case of high-power TWT amplifiers used for some small video uplinking terminals, forced-air cooling is required.

The feed horns are often protected by a plastic radome material to keep out moisture, which would interfere with the signals and corrode the feed. Care is taken to hermetically seal the outdoor unit to prevent damaging ingress of moisture (see Figure 2.3).

The mounting system allows the attachment of the antenna/feed system to a suitable base. The mount must be rigid for operation, but must permit smooth rotation of the antenna about one or more axes for alignment with the satellite. In addition to that, the feed must be rotatable for proper alignment with the polarization of the satellite signal. This requires the ability of the feed to rotate about its longitudinal axis, since linear polarization is used.

The various feed/RF combinations provide different means for rotating the feed horn. Some permit rotation of the feed horn with respect to the electronics package; others rotate the feed and electronics package as one integral unit. Separate rotation of the feed with respect to the electronics

Visor to keep rain, ice, and birds away from feed

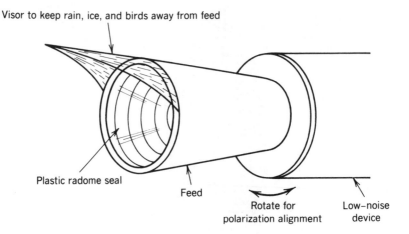

Plastic radome seal

Feed

Rotate for
polarization alignment

Low–noise
device

Figure 2.3. Sealing the feed opening.

box may require resealing of the enclosure in the field. Combined rotation eliminates this problem but necessitates a more formidable attachment for the electronics box.

The mounts usually provide for some form of screw-thread adjustment for fine elevation alignment with a subsequent lock-down method for operation. Mount materials are properly weatherized steel or possibly, aluminum. The hardware items should be made from stainless steel to minimize corrosion.

There are many suitable mounting methods for these small antennas. Roof mounts come in penetrating and nonpenetrating varieties. Ground mounts consist of a single steel pole secured with concrete in a properly prepared hole in the ground, or tripods (and similar arrangements), anchored to prepared footings. In urban areas a roof or wall mount may be mandatory. Roofs generally have low load ratings per square foot, and therefore, the antenna loads, which include the weight of the antenna system, the wind loading, snow, and ice, must be spread over a wider area.

Care must be taken to assure the mechanical stability of the mount in high-wind conditions to keep the antenna pointed at the satellite. A number of the vendors specifically mentioned the wind speeds in which their antennas can operate with little or no degradation and the wind speeds that the system will survive. The latter is an important parameter to determine possible liability risk by the antenna coming loose and causing injury. In the mid-1980s the Communications Center conducted a study for an insurance company. We did not find any antennas that blew loose after proper installation, but we did encounter several cases when an antenna blew or rolled off a roof during installation.

The other component of the terminal system is the indoor unit, which is generally connected to the outdoor unit by coaxial cables. These cables carry some convenient intermediate frequency (IF), generally in the ultrahigh-frequency range, for microterminals. Additional cables may be needed to provide electric power and/or control signals to the outdoor unit. Figure 2.4 shows a multiwire cable that may be used. For television station use the rotor may be needed to receive signals from different satellites at different times. In most microterminal applications, the antenna is pointed at only one satellite and hence the rotor wires are not needed.

2.1.2 The Indoor Unit

The indoor unit usually operates in a controlled environment. A block diagram of a typical small earth terminal is shown in Figure 2.5.

The indoor unit contains the data, voice or video receivers, modems, codecs, control equipment, signal processing equipment, user interface equipment, and possibly power conditioning equipment for both the indoor and outdoor units. Prime power is generally 110 V ac. Most of the outdoor units require only low-voltage dc power, thus avoiding the need for 110 V ac power lines to the antenna. All outdoor equipment should be grounded as a protection against lightning and voltage buildup. Most low-noise receiving equipment operates in the range 15–20 V. A vacuum-tube power amplifier generally requires high voltage (over 100 V dc), whereas a 1- to 10-W solid-state amplifier may be supplied with less than 28 V (ac or dc).

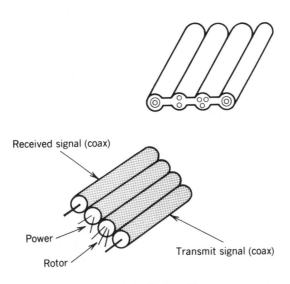

Figure 2.4. Single cable.

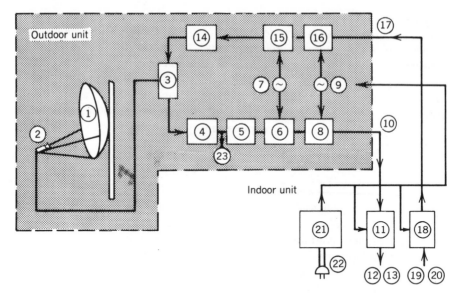

1 Antenna reflector
2 Antenna feed
3 Orthomode transducer
4 Transmit rejection filter
5 Low noise amplifier*
6 Frequency agile down converter*
7 Frequency synthesizer local oscillator**
8 Fixed frequency down converter**
9 Fixed local oscillator**
10 Intermediate frequency output
11 Modem codec
12 Interface (RS-232, for instance)
13 Output to end user

14 Power amplifier
15 Frequency agile up converter
16 Fixed frequency up converter**
17 Intermediate frequency input
18 Modem codec
19 Interface (RS-232, for instance)
20 Input from user
21 Power supply
22 Local AC supply (110 VAC, 60 Hz)
23 To video receiver (for one-way teleconferencing)

*These functions may be combined in a low
 noise converter (LNC)
**These functions may be located in the indoor
 unit

Figure 2.5. Microterminal block diagram.

2.1.3 Installation and Maintenance

The installation phase is actually preceded by several tasks which may be time consuming. First, permission from the landlord to erect an antenna must be obtained. There may also be a need to get permission from the local authorities (zoning, planning, historic district, etc.) especially if the antenna is larger than 1.8 m. For C-band services a site survey, services of a frequency coordinator, and filing with the Federal Communications Commission (FCC) are needed. At K_u-band, type approvals in a satellite exclusive band (except near four U.S. military installations) simplifies the last part.

Installation of the microterminal is usually performed by semiskilled technicians (the Equatorial antenna is an exception, in that it is intended to be installable by reasonably handy laypersons). Installation times range from less than an hour (for temporary or truck-mounted systems), to 1 or 1½ days for permanent sites. Actual installation times and costs can, of course, vary considerably depending on the individual locations. In some locations, deicing equipment may be required to keep the antenna and feed system free of ice in the winter time.

In keeping with the Communications Center's goal of having "hands-on" experience so that it can advise its clients authoritatively, we had a 1.2-m K_u-band terminal installed in 1984.

The 1.2-m receive-only antenna at the Communications Center required two people 4 hours to install. At least 2 labor-hours were spent in tracing the grounding system (see Figure 2.1) through the building and eventually through an underground power cable to a ground at the electric pole. The building also had an extensive series of lightning rod grounds.

The actual time included digging a 4-ft-deep hole (with a gasoline-powered earth auger), mixing and pouring concrete, setting the pole, aiming the antenna (5 minutes), running 26 m (85 ft) of coaxial cable, burying 10 m (33 ft) of the cable, boring entry holes into the building, running ground wires, connecting the 12-GHz receiver, and so on.

Pointing a small antenna at the satellite would seem to be an easy task considering that a 1.2-m antenna has a half-power (3 dB) beamwidth of 1.5° at 12 GHz. In reality, the pointing is fairly critical. The installation (or maintenance) crew should have a computer printout of the azimuth (compass pointing angle from *true* north) and elevation (the angle from the horizontal) for the stations location. The elevation is easily found with an indicating level and the antenna should be accurately pointed in this axis. A pocket compass is generally not accurate because it measures angles from the *local magnetic* north. The error is the difference between the *true* and *magnetic* north. This difference, called deviation, may be found on topographic maps or obtained from the local airport. The difference can be many times the 1.5° beamwidth calculated above. A "smart" computer printout would contain a corrected compass reading for antenna pointing. After roughly pointing the antenna in azimuth the antenna should be slowly rotated in azimuth until a signal is found in a preselected transponder. After verifying that the signal is from the right satellite (yes, there are others!), the azimuth and elevation adjustments should be varied to bring the signal level to the magnitude encountered in nearby installations. If the signal is too weak, the antenna may be seeing the satellite via one of its sidelobes, the polarization may be misadjusted, or it may be the wrong satellite.

If the earth station antenna has a beamwidth of 0.4° or less (4.5 m or larger at 12 GHz) and it does not have a satellite tracking capability, it may be necessary to find if the satellite is at its nominal position or if it has drifted slightly off nominal. See Chapter 13 for a further discussion of antenna pointing.

Maintenance requires a higher skill level. In some systems no field maintenance is provided on the indoor units. An overnight courier service (e.g., Federal Express, Purolator, UPS, etc.) delivers a replacement controller box in the morning and takes the original box back to the maintenance depot for rework.

2.1.4 Antenna Diameter

Two K_u-band typical sizes have evolved. These are 1.2-m (4-ft), and 1.8-m (6-ft) diameters. These sizes are possible because of the higher (than C-band) power of the domestic K_u-band satellites in the United States, and the limited traffic requirements of this new class of station. The use of a spread spectrum at C-band permits 60-cm one-way and 60 cm $\times$ 1.2 m (2 ft $\times$ 4 ft) for two-way low-speed data.

Although some of the K_u-band terminals are 1.2 m in diameter, the Communications Center feels that wherever possible, the larger 1.8-m antenna is to be desired. This provides an additional 3 dB in receiving and transmitting gain. This may be used to make the station more robust or to accommodate future growth. The space system operating economics are inversely affected by the microterminal diameter. The 1.8-m antenna also provides compatibility with the 2° separation between domestic satellites. This is particularly true for transmit stations. We further predict that, in time, the 1.8-m antennas may eventually be retrofitted to 2.8–3 m, due to the limitations and economics of the space segment. The negative aspects of the larger antennas will be the higher installation costs, the more difficult pointing, and the need for a bigger antenna mount. In many cases, a new zoning hearing may be needed. The spread-spectrum C-band stations use the 60-cm receive-only and 60 cm $\times$ 120 cm receive/transmit antennas.

2.1.5 Microterminal Locations and Restrictions

These stations are typically located on a customer's premises, a short cable run from the customer's computers, PBX equipment, or point-of-sale terminals. As such there is no need for a local telephone company to provide any service. This is a form of total bypass of these telephone companies. Unlike the United States, the PTTs of other nations may vigorously oppose microterminals for this reason. Industry, on the other hand, will argue for an unfettered ability to communicate. We expect that many forms of compromise will evolve.

When transborder telecommunications is involved, it gets even more complex. Some transborder business communications are already in place (United States to Mexico and Canada) using U.S., Mexican, and Canadian common carriers. Extensive noncarrier transborder networks have yet to be established. In addition to the existing carriers, this topic is of interest to organizations such as Intelsat, Inmarsat, the Eutelsat, Arabsat, Palapa (Southeast Asia), multinational companies/organizations, and the new private regional/multiregional satellite networks.

2.2 FREQUENCY BANDS

Microterminal services may use the fixed satellite service (FSS) frequency bands shown in Table 2.1 (see also Chapter 12). The S-band is not used for satellite services in the United States because of terrestrial interference and the narrow bandwidth of the allocation.

The C-band is widely used for other domestic services (such as distribution of television to cable-TV earth stations, network television, etc.) This band is also used extensively for point-to-point terrestrial microwave networks. Conventional microterminal applications would be difficult to implement at C-band because of the interference. Spread-spectrum multiple access, on the other hand, is relatively immune to this interference. Contel's Equatorial Communications of Mountain View, California, has been successful in operating in this band.

The X-band is reserved for government uses. In Europe and the Atlantic,

TABLE 2.1   Fixed Satellite Service Bands for North America

Letter Band	Primary Telecommunications Satellite Frequencies (GHz)		Short Name
	Earth to Space (Uplink)	Space to Earth (Downlink)	
L	1.6265 – 1.6605	1.530 – 1.559	Mobile
S	2.655 – 2.690	2.500 – 2.655	2.5
C	5.925 – 6.425	3.700 – 4.200	6/4
X	7.900 – 8.400	7.250 – 7.750[a]	8/7
K_u	14.000 – 14.500	10.950 – 11.200[b] 11.450 – 11.700[b]	14.11
K_u	14.000 – 14.500 17.300 – 18.100	11.700 – 12.200 12.200 – 12.700	14/12 DBS[c]
K_a	27.500 – 30.000	17.700 – 20.200	30/20

[a]Government only.
[b]International (e.g., Intelsat) only.
[c]DBS, direct broadcast satellite.

it is used by the French *Telecom-1*, U.S. *DSCS*, British *Skynet*, and several series of Soviet satellites.

The K_u-band (14/11 GHz) is used only for international satellite services in the United States. The only earth stations using microterminals in this band at this time are in the Intelnet service. Terrestrial services use the 11-GHz band in the United States for point-to-point microwave.

The primary frequency band of interest is the second K_u-band (14/12 GHz), which is being used for domestic satellite services. The direct broadcast satellite frequency bands (17/12 GHz) are not presently in use. It is expected that someday these bands may be used for business services in North America. The K_a-band (at 30/20 GHz) is not in service at this time in the United States, but both Japan and Europe have service plans at 30/20 GHz.

The mobile satellite service (MSS) could also provide microterminal services. The inherent limitations are bandwidth (2.5–5 kHz per channel) and the availability of frequencies in the L-band (1.5 and 1.6 GHz). These services may be limited by *Inmarsat* and Soviet satellites with global coverages.

2.3 PRIVATE NETWORKS

Most microterminal voice and data networks presently envisioned are for internal use by only one corporation and its customers. This is done, in part, to avoid regulatory problems that might arise through the interconnection of private and public switched networks.

The present exceptions are the microterminal services offered by AT&T Communications and those offered by public packet-switched networks (Tymstar and Telenet). While presently structured as a point-to-point service, this could easily be interconnected with a public switched network, thereby providing AT&T with the opportunity to bypass completely the local exchange carriers (which were part of the Bell system before divestiture).

Because most of the networks are self-contained, there is a feeling that they do not have to be standardized with other network interfaces. Although this would seem to lead to a wide variety of operating standards, most of the manufacturers of the equipment prefer to produce a product that is compatible with the existing telephone and data systems in the United States and the world. Standardized connectors (e.g., the modular telephone and RS-232 connectors; see Figures 2.6 and 2.7) are widespread. Facsimile networks generally follow the CCITT standards, and so on.

As networks mature, we expect that more protocol levels (see Chapter 11) will be added to what are now quite simple networks that involve little switching. We note the emergence of the shared telehub where one hub station serves many private networks. These eventually will interconnect, first with one another and later to the public networks.

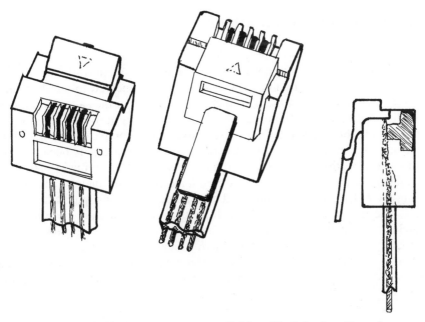

Variations of this connector are available with 2 to 8 active pins.

Figure 2.6. Modular telephone connector.

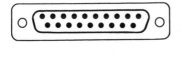

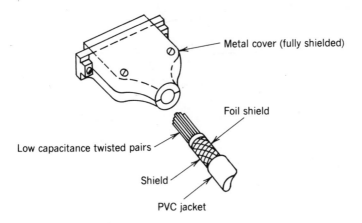

Metal cover (fully shielded)

Foil shield

Low capacitance twisted pairs

Shield

PVC jacket

Figure 2.7. RS-232 connectors.

2.4 NETWORK CONFIGURATIONS

Most of these networks employ a single large earth station (referred to as a hub) and operate on a star network basis (Figure 2.8). All traffic flows from a small terminal into the hub. If it needs to be connected to another small station, it is processed at the hub, then retransmitted to the same (or another) satellite and thence to the other small station. Mesh networks (Figure 2.9) are possible, but tend to be more expensive, more complex to operate, and consume the space and earth segments more rapidly. This is because each of the small stations must have sufficient power and receiving sensitivity to communicate among themselves. A pseudomesh network may be constructed using the hub as a transit point. All communications must pass through the hub. The use of a hub station, with its large antenna gain, permits more efficient use of the space segment and the ability to make up for some of the lack of gain in the small stations. Although the hub station may be expensive, its cost is divided many ways (by the number of small stations) and becomes a number far smaller than the enhancements that will be required

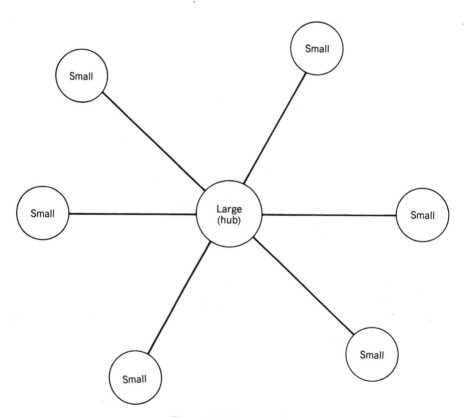

Figure 2.8. Star network.

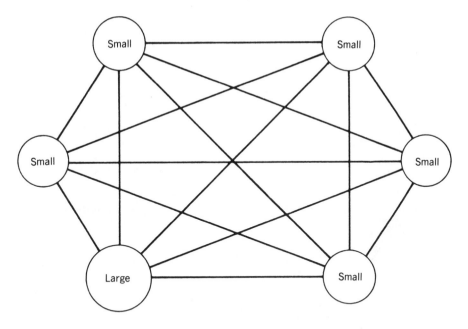

Figure 2.9. Mesh network.

in each of the small stations to make them fully mesh compatible. All the major microterminal networks are stars.

2.5   MULTIPLE-ACCESS AND MODULATION

There are many ways to organize a microterminal network. Most network designers try to optimize the network organization to conserve satellite re- sources as well as satisfying the customer's requirements. Network orga- nization requires some form of access control for the usually fairly large number of microterminals comprising the network. Generally, the designer will choose a modulation scheme in conjunction with a multiple-access scheme. Both the multiple-access and modulation schemes are likely to be different for the link from the microterminal to the hub and from the hub to the microterminal. Figure 2.10 shows some typical multiple-access and mod- ulation schemes. Multiple-access schemes fall into a number of classes, each with some distinct advantages and disadvantages.

2.5.1   Spread-Spectrum Multiple Access

Spread-spectrum multiple access (SSMA) relies basically on the code within a pseudorandom spreading sequence to provide separation between multiple

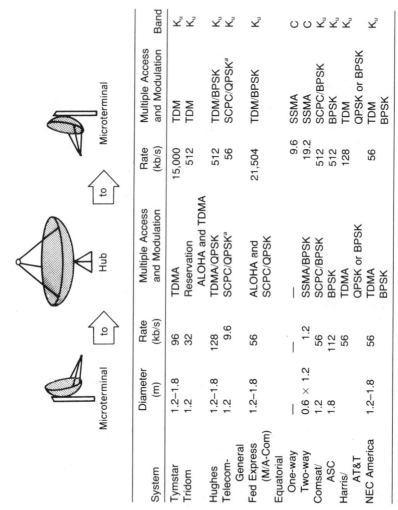

Figure 2.10. Network operating methods.

System	Diameter (m)	Rate (kb/s)	Multiple Access and Modulation	Rate (kb/s)	Multiple Access and Modulation	Band
Tymstar	1.2–1.8	96	TDMA	15,000	TDM	K_u
Tridom	1.2	32	Reservation ALOHA and TDMA	512	TDM	K_u
Hughes	1.2–1.8	128	TDMA/QPSK	512	TDM/BPSK	K_u
Telecom-General	1.2	9.6	SCPC/QPSK[a]	56	SCPC/QPSK[a]	K_u
Fed Express (M/A-Com)	1.2–1.8	56	ALOHA and SCPC/QPSK	21,504	TDM/BPSK	K_u
Equatorial						
One-way	—	—	—	9.6	SSMA	C
Two-way	0.6 × 1.2	1.2	SSMA/BPSK	19.2	SSMA	C
Comsat/ASC	1.2	56	SCPC/BPSK	512	SCPC/BPSK	K_u
Harris/AT&T	1.8	112	BPSK	512	BPSK	K_u
		56	TDMA	128	TDM	K_u
			QPSK or BPSK		QPSK or BPSK	
NEC America	1.2–1.8	56	TDMA BPSK	56	TDM BPSK	K_u

[a]May use TDM or a form of spread spectrum.

Microterminal → to → Hub → to → Microterminal

signals which are transmitted simultaneously through the satellite. The modulation of the carrier is normally BPSK. Spread-spectrum multiple access is quite bandwidth inefficient, but it is highly interference resistant. It is used effectively by Equatorial at C-band.

2.5.2 Single Channel per Carrier

Single channel per carrier (SCPC) is essentially a frequency-division multiple-access scheme. Each microterminal transmits a message on a fixed frequency. The selection of the transmit frequency can be permanently fixed (i.e., preassigned) or selectable by the microterminal or assignable to the microterminal from the hub station on a message-by-message basis. Modulation schemes can be BPSK or QPSK, depending on the available bandwidth, power, and coding.

2.5.3 TDM and TDMA

Time-division multiplex (TDM) is most often used in the link from the hub to the microterminals. TDM means that the transmission from the hub is a continuous stream containing information to the individual stations in a time-sequenced set of message bursts.

TDMA (time-division multiple access) is a time-sequenced set of messages originating from different microterminals that transmit on an assigned frequency in time slots that are controlled according to some preestablished algorithm. Sometimes the transmissions are assigned according to reservations made in advance on a separate channel.

Reservations are frequently made on a channel according to an Aloha algorithm which basically permits random access in time with the hope that accesses do not collide at the satellite and a means is provided for retrying the collided transmissions.

In view of the asymmetry of the links from microterminal to hub and hub to microterminal, the data rates in the two directions generally also vary. This is discussed further in Chapter 13. Figure 2.10 is a summary of some of the network multiple access and modulation arrangements.

2.6 THE CONTROLLER

A typical microterminal contains a controller that contains connectors for synchronous and asynchronous digital connections to computers, word processors, and so on. There may also be connections for voice (telephone) and possibly analog video. The size of a typical controller is illustrated in Figure 2.11.

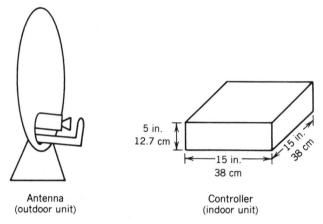

Antenna
(outdoor unit)

Controller
(indoor unit)

Figure 2.11. Station controller: microterminal elements.

2.7 PRICES

The microterminals are characterized by per station prices (in large quantities) at previously unheard of low levels. When Federal Express indicated that they were interested in buying up to 50,000 earth terminals from a single manufacturer at $10,000 each (installed), there were many skeptics. Considering that the order would amount to $500 million, many organizations decided to see what could be done if large quantities of these stations were ordered.

Figures 2.12 and 2.13 illustrate the economy of scale and the sensitivity of the price to the demand. In Figure 2.12 the price to produce successive units drops as the quantity produced increases. Not shown in the figure are potential steps in the curve which come about when the assurance of large quantities of future orders encourages a complete product redesign (for low-cost manufacturability), the construction of special tools and jigs, and so on. Mid-1987 prices are in the range $6000–$16,500. The cost reduction with quantity situation has occurred in both the home earth station (see Figure 2.14) and the videocassette recorder (Figure 2.15) markets.

As long as earth stations for television reception were priced in the range of $25,000–$100,000, the market was very limited. When the price dropped below $20,000, it became attractive for cable television systems to acquire stations. Substantial growth in the home market awaited the $10,000 threshold and still later the $2000 line. Typical television receive-only (C-band) stations for home reception in the United States sold for $1500–$2000 in mid 1987. There are now over 1 million such stations. An external influence (scrambling) has seriously altered the market demand.

Videocassette recorders hovered at the $600 range for several years. After

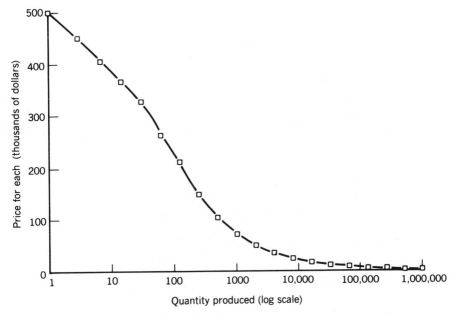

Figure 2.12. Economy of scale.

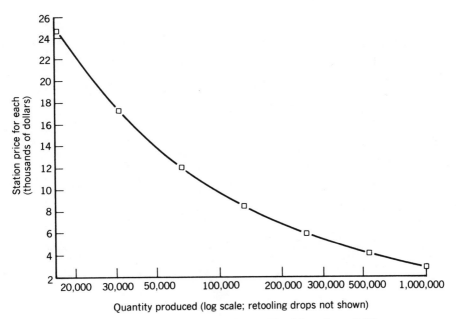

Figure 2.13. Future economy of scale (detail from figure 2.12).

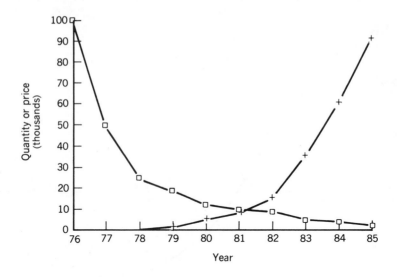

□ Price   + Total number in use
  (in thousands of dollars)   (in ten thousands)

Note: Prescrambling

Figure 2.14. Home earth stations.

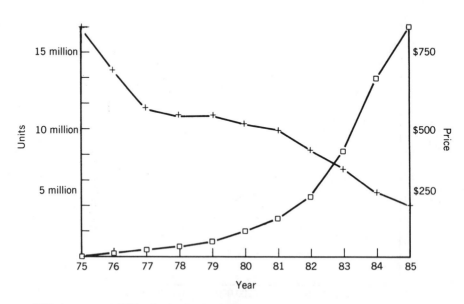

□ VCRs in use   + VCR price (each, at retail)

Figure 2.15. Video cassette recorders.

the $500 price was reached, sales began to climb. The quantity produced has brought the price of these machines down into the $250 range.

Although these have been consumer products, it is expected that the availability of truly low cost ($9000–$15,000 installed) business earth stations offering highly reliable cost-stable telecommunications may prove irresistible to corporate managers. The price range is broad due to the wide variety of speeds and capabilities available.

2.8 TYPICAL SPECIFICATIONS

Table 2.2 shows the basic specifications of typical microterminals. For further information, consult the technical sections.

TABLE 2.2   Typical Microterminal Specifications

Antenna diameter	1.2, 1.8, 2.4 m
Receive frequency	3.7–4.2 GHz or 11.7–12.2 GHz
Transmit frequency	5.925–6.425 GHz or 14.0–14.5 GHz
Typical transmitter power rating	1–10 W
Bit rate	1200–9600 b/s or 1.544 Mb/s

3

MICROTERMINAL APPLICATIONS

3.1 ONE-WAY BUSINESS SERVICES

3.1.1 Video (One-Way)

Business television is defined as nonentertainment television service. We have excluded the conventional television network and cable television services distributed by satellite. Business television makes use of the technology that was originally developed for the unsuccessful United Satellite Communications Incorporated (USCI), which pioneered the use of small antennas and relatively inexpensive low-noise converters and receivers intended for direct reception of entertainment TV in the home. Whereas USCI used 1.2-m (4-ft) antennas, most of the business video users use a 1.8-m (6-ft) antenna. The typical noise temperature of the low-noise converter is approximately 225 K.

The most exciting part of the microterminal situation is the new uses of the small terminals. Many of the applications were not considered seriously a few years ago. The success of the Private Satellite Network (which handles business television) has been a surprise, as most of the teleconferencing business had failed to develop for other service providers. The difference is that the Private Satellite Network (PSN) has developed a unique opportunity that provides facilities for internal corporate uses (such as training, merchandising, daily meetings, etc.). This is different from either the wide-screen presentations that connected one large conference to another or the businessperson-to-businessperson video phone calls. PSN can provide the programming and the telecommunications facilities. At the present time, PSN has over 1500 terminals installed. These are for one-way business video using full- or half-transponder B-MAC television.

PSN's business television is used in such applications as the J.C. Penney retail stores and Merrill Lynch (a stockbroker) for internal teleconferencing on a point-to-multipoint basis. Teleconferencing (point to point) may also be done between business offices and meeting rooms.

Most of the point-to-multipoint services use either a full transponder or a major fraction of a wideband transponder. Scrambling is often used to protect the content of the transmission against eavesdropping by competitors. The scrambling may take several forms. The most common varieties use a secure audio scrambling system and a less secure video service. Generally speaking, if the audio is denied to an eavesdropper, the video value is substantially lowered. In some cases, multiplexed analog component (MAC) video is being introduced to enhance the quality of the signal. The MAC systems have the capability of handling video, encrypted TV-audio, plus nonrelated voice or data channels; all of these channels (except video) are digital.

If an organization has T1 (1.544-Mb/s) microterminals, business videos may be accommodated using compression. Video compression is expensive. It takes a standard television picture and reduces it to a full or fractional T1 carrier (see Table 3.1). The fractional T1 video is attractive since the remaining T1 channel capacity can be used simultaneously for other voice and data services.

Still lower bit rates can be used for video, but the quality degrades at lower speed even with advanced compression techniques. Rapid motion can cause the picture to break up or get fuzzy (depending on the encoding method) in the region of the motion. As shown in Table 3.1, some equipment is avail-

TABLE 3.1 Teleconferencing Equipment

Organization	Location	Description	kb/s	
Avelex	Silver Spring, Md.	256 × 240 pixels, B&W motion video	56	(T0)
		Color full motion	112–512	
Colorado Video	Boulder, Colo.	Slow-scan TV	(1–8 kHz)	
Compression Labs, Inc.	San Jose, Calif.	Full color and full motion	1,544	(T1)
		Color frames	56	(T0)
		"T1 quality"	768	
		Reduced data	512	
NEC America, Inc.	Fairfax, Va.	NTSC	1,544	(T1)
		NTSC	6,312	(T2)
		128 lines × 128 pixels	56	(T0)
		Digital NTSC	45,000	(T3)

able at 56-kb/s rates. Even lower rates can be used if "freeze frames" are acceptable. In this case the video consists of a series of "snapshots" transmitted periodically. The equivalent of a voice telephone channel (3–4 kHz) can be used for this lower extreme.

Video signals have even more redundancy than voice. The various horizontal and vertical synchronizing signals are, by definition, repetitive. Much of the video information is unchanging from scan line to scan line and even within a line. Between frames, there may be very little (often no) change at the same point on a line. In the case of a typical teleconference frame, the background remains fixed and only the speakers' lips move. The changes in information content from frame to frame may be very small. By sending only the part of the video frame that is altered (the "delta" in delta modulation), the bandwidth may be reduced substantially.

When the scene changes totally (by switching from one camera to another or moving a camera or background), the video compressor must cope with a sudden flood of new information. This may be done by initially transmitting only the broad area information and then sending progressively more detailed material in successive frames to "paint" the rest of the picture. This may be done so rapidly that the eye is unaware that it is happening. Error detection and concealment methods may be used to reduce the bit rate further and make the signal more resistent to noise and interference.

Whereas a fully encoded network-quality video signal may require 45–90 Mb/s, compression may reduce the rate to 9–25 Mb/s. Teleconferencing is less demanding and may routinely be accomplished at 1.544 Mb/s (T1). Still-slower-speed systems are available to 56 kb/s.

It is expected that eventually internal corporate networks may appear to provide the same services, but using lower-bit-rate equipment (e.g., T1 or fractional T1) video terminals. Since the T1 microterminal is insensitive to the type of communications being passed through the terminal, it may be used simultaneously for fractional T1 video, data, and voice services. The mixture may vary throughout the business day.

3.1.2   Data (One-Way)

Stocks, Bonds, and Commodities.   One of the earliest applications of this type of technology was for the distribution of stock and commodity market information via the Equatorial system. All of the transactions of the New York and American Stock Exchanges are transmitted by satellite to small microterminals. These terminals are typically 60 cm (2 ft) in diameter and operate on a one-way (receiver only) basis at 4 GHz. The Chicago Board of Trade and the Chicago Mercantile Exchange trade in commodity futures. These include such items as grains, meats, and other foodstuffs. This information is sent to various brokerage and sales organizations across the United States by satellite and microterminals. In this case a private leased line (Chicago to near San Francisco) carries the data to the hub, from which it is transmitted to *Westar IV* (or *Galaxy III*), and thence to microterminals in the coverage area of the satellite.

Price Lists and Inventory. During times of rapid inflation or competitive pressures, the price and availability of goods may be changing rapidly. A high-speed telecommunications system to keep track of these changes in prices and inventory can be used as a means of controlling the amount of materials that are in stock and locating goods that are in transit or in storage at other locations. In this way, the efficiency (therefore, the profitability) of a manufacturing and distributing organization may be improved substantially while giving the customer better service. Savings may be achieved by reducing the inventory (and thus the investment) if the whereabouts of goods are known so that they can be found when needed. If a merchant cannot find the goods, they cannot be sold. When they are found, they may later have to be disposed of at a sale price. With management vitally interested in cash flow, a microterminal network could pay for itself just by reducing the inventory excesses.

Now that hard disk and semiconductor ("hard card") memories are readily available for small micro- and minicomputers, it becomes feasible to transmit inventory records and price changes to outlying locations at night (when there may be little other traffic) and store the list at the point of sale (such as in the checkout computers of grocery stores). During the day (when the capacity may be in higher demand for other transmissions), only the changes of substantial consequence are sent to the outlying locations to keep their data current.

Retail Sales. Many retail outlets in the United States use computerized cash registers (or point-of-sale terminals). Many of these include bar code readers. These keep running tabulations on the sales and inventory within the store. Tape records may be sent (via telephone lines or courier service) to a regional headquarters where sales trends and replacement orders for goods sold may be processed. When manual handling of the tapes is required, many days may pass before the information is processed and trends detected.

In some retail companies, 40–50% of the annual sales are conducted in November and December (Christmas time). The ability to detect trends quickly and order new goods during this short spurt of sales can be critical. Conversely, the cancellation of orders for goods that are not selling is also important. Through the use of a two-way microterminal, the sales records may be transmitted automatically to a central facility either on an hourly or daily basis (during the nonbusy hours at night) so that tight inventory control may be maintained and profitability maximized. Inventory that gets "lost" in these two months may remain on the books until the next Christmas.

Even in the case of a steadier business traffic (e.g., fast-food franchises), a microterminal network that collects data on the sales levels of each menu item from each store can alert the local commissary to increase the hamburgers for store 1234 by 10% and cut the ice cream back by 25%. Considering the rapid turnover of some store managers and their youth, this may be more than just a convenience.

3.1.3 Audio (One-Way)

Generally speaking, the smallest terminals (particularly those using spread spectrum) do not handle voice service on a two-way basis. In general, this is because of the uplink power requirements. There is no reason why this cannot be overcome when improved technology is available.

Using digital (rather than analog) voice, some of the newer digital voice processing equipment voice may be handled on a channel substantially lower than the 32- to 64-kb/s per voice channel currently in widespread use. Table 3.2 shows various types of voice processing techniques. Figure 3.1 shows typical processing delays with the electronics of the mid-1980s. With time (and faster electronics) these delays will be reduced. 16-kb/s voice codecs are commercially available from several sources.

When many voice lines are encountered, the T1 services may be used economically. These encode the voice at 32 to 64 kb/s and combine them into a 1.544-Mb/s bit stream. Twenty-four 64-kb/s or forty 32-kb/s half circuits can be derived from a T1. The economics between T1 and individual voice line costs cross over at about 10 voice circuits. Naturally, T1 requires a wider-frequency band and a more powerful earth station with a larger-diameter antenna than for 56-kb/s service. Typical antenna diameters may be in the 2.4-m (and up) range. Most T1 services are used on a point-to-point basis; however, it is conceivable that T1 could also be used on a point-to-multipoint basis if the hub provides the function of a baseband switch.

Why should a 4-kHz voice signal require 64,000 b/s? That is a question that has bothered research laboratories for some time and why those labs have spent time trying to reduce the number of bits needed to transmit both

TABLE 3.2 Voice Processing Techniques

Transmission Method[a]	Coder Type	Typical Data Rate (kb/s)	Relative Quality
CH-V	Source	2	Poor
LPC	Source	2	Poor
	Source	4	Fair–good
DM	Waveform	16	Good
	Waveform	32	Good–very good
ADPCM	Waveform	16	Good–very good
	Waveform	32	Very good
Wideband ADPCM	Waveform	64	Very good
PCM	Waveform	64	Very good

[a]V, vocoder; LPC, linear predictive coder; PCM, pulse-code modulation; ADPCM, adaptive differential PCM; DPCM, differential PCM; DM, delta modulation; CH-V, channel vocoder.

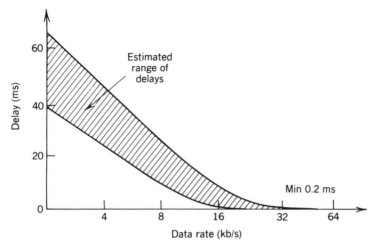

Figure 3.1. Processing delay times. (Reprinted with permission of the *GEC Journal of Research*, published by The General Electric Company plc of England.)

voice and video information. The benefits can include reduced earth station size or added services. In fact, there are systems in place that operate with only a fraction of the traditional 64-kb/s encoding rate, and several processes are available for compressing data in both voice and video services.

An audio signal should be sampled at least twice when it is converted from analog to digital form to permit reconstitution back into its analog state. Assuming that the highest audio frequency in a telephone conversation is 4000 Hz, the sampling rate must be 8 kHz (or greater). The 8000 samples per second may be sent through a quantizer which measures the signal (to 1 part in 256) and produces a 7-bit digital code. Seven bits times 8000 samples per second results in 56 kb/s.

In the transmission process, noise might slip in and change one of the bits from a one to a zero, or vice versa. An eighth bit is added so that the number of ones in the quantized signal is always an even (or odd) number. This eighth bit is called a parity bit and is used to determine if an error occurred in the transmission. This works on the principle that the odds of making two errors of the same binary sense is very small. Eight thousand samples per second multiplied by 8 bits per sample produces the familiar 64 kb/s.

To cut the bit rate in half (to 32 kb/s), the sampling rate could be halved (but the highest audio frequency would be 2000 Hz). This is unacceptable for normal telephone service. A more fruitful method is to examine the signal. Many sounds are repeated in conversation so it often is possible to predict on the basis of previous samples what may be in the next sample.

By using such techniques as linear predictive coding (LPC), the repetitious material may be removed from the conversation while still retaining the basic information content. With LPC, repeated sounds are deleted from the transmission end and replaced with a code, reducing the required number of bits. At the receiving end, the code is transformed back into the deleted sounds so that the conversation makes sense.

Delta modulation (DM) is another method to reduce the number of bits per sample, cutting the total bit rate. This system measures the difference (the delta) between one sample and the next and has a much smaller dynamic range than sending the absolute amplitude of each sample.

SBS pioneered the widespread use of 32 kb/s and has received high marks from various consumer organizations on the clarity of its service. AT&T and others have recently adopted 32 kb/s as an alternative to the long-standing 64-kb/s service. This means that two voice channels can occupy a single 64-kb/s channel. There may be applications at still lower bit rates such as 4 kb/s for various recorded message services when the information is primarily one way.

There are other one-way applications where it is not vital to recognize the speaker's identity, but the information content is very important. Examples include economic data, news, and air traffic control. In these cases the bandwidth reduction that occurs due to the data compression may be used to extend the range of the service, reduce the receiver cost, or increase the capacity of the telecommunications facility.

News, Radio, and Other Audio Communications. Business audio includes motivational music and advertisements (such as used in grocery stores to promote products). These are point-to-multipoint services which may be handled on either a digital or an analog basis. At the present time, the principal methods used in the United States are single channel per carrier (SCPC) or time-division multiplex (TDM). In SCPC, the transponder is frequency divided and individual carriers are used for each audio channel. In some cases the audio channels are transmitted in some form of stereo modulation.

In the case of TDM, the transponder is time-shared among many radio or audio networks. The typical receiving antenna for the services is 3 m in diameter. Because of the increasing cost of terrestrial telephone lines, the poor quality (typically, 3–4 kHz), and the difficulties in long-distance analog transmissions, many business audio networks in the United States have converted to satellite distribution. Eventually these services may use digital microterminals.

3.2 TWO-WAY BUSINESS SERVICES (EXCLUDING VOICE)

3.2.1 Video: Two-Way Teleconferencing

As indicated above, point-to-multipoint television services for businesses tend to use full transponders (or 15–25 MHz of a wideband transponder). Two-way services (as in the case of a true teleconference) generally require at least two video channels (one from each end).

Through the various video bandwidth compression techniques, highly acceptable conference room-to-conference room (or office-to-office) video may be produced at one-third, one-half, and full T1 rates. Since these are digital transmissions, they are very difficult to intercept, and unless the same type of equipment is used by the intercepter, they are hard to decipher. Encryption of the signals is also readily possible and inherent in some systems.

In addition to corporation internal communication (news and so on) normal business transactions, teleconferencing may be used for sales and training purposes. Hewlett-Packard, J.C. Penney, Wang, Federal Express, IBM, and the U.S. military make use of this type of training. It saves substantially on travel, and if extensive training is required, family relocations. Through the use of a two-way video facility the students are permitted to ask questions of the instructor and to demonstrate their work remotely. In many instances, however, the return link may consist of a terrestrial audio telephone line (often a toll-free number).

3.2.2 Data (Two-Way)

The insurance industry in the United States has become very complex. Insurance rates are established on the basis of past history of a large number of insurance policies. Different models and makes of automobiles, for instance, have different risk rates and insurance premiums. In some instances, this information is resident at the home office of the insurance company. The individual branch offices and agencies are getting two-way data microterminals that can access the home office data base, thereby permitting immediate rating of the risk and computation of the premium. One of the problems in insurance is that people change their mind or consult competitors if given enough time. By computing the rates immediately (while the client is in the office), the insurance agent has a much higher probability of making the sale.

Point of Sale. At each point of sale two important transactions take place. The first is the actual sale, which involves the current price and availability of the merchandise. We discussed price lists and inventory control in Section 3.1.2. The second part deals with getting paid for the product or service, which involves verification of the customer's credit. This takes place whether the purchase is paid for by check or by a credit card.

Because of the risks to the merchant in accepting charges against a stolen or counterfeited credit card or against an account that is substantially overdrawn, credit verification is often required. At least one chain of gasoline stations is considering using small microterminals at its major stations to verify the credit cards of its customers.

Automatic bank teller machines (ATMs) are another potential market for microterminals. The condition of the customer's bank account must be checked before cash is issued by the ATM. Since the ATMs of many U.S.

regional banks are now tied into large area networks, the bank's computer may be hundreds of miles away from the ATM.

Financial Reconciliations. The U.S. Federal Reserve Board requires that banks in the United States reconcile their accounts receivable and payable each day. Considering the huge amounts involved in transfers between major banks and the time value of money (particularly when the prime interest rate is 10% or more), there is a great incentive to reconcile these accounts as rapidly as possible and with the greatest reliability.

In the state of California, several of the banks have discovered that even as a result of relatively mild earthquakes the terrestrial telephone network between banks (particularly for the bulk reconciliations) can be disturbed even if dedicated leased lines are used. In at least one instance, one southern California bank operated for several days while the terrestrial network was being repaired using their Los Angeles-to-San Francisco satellite link to make these reconciliations.

Transoceanic Service. Tables 3.3 and 3.4 show the Intelsat and private system international earth station standards. As can be seen by Table 3.3, the increasingly smaller earth terminals require more power or bandwidth from the satellite, and therefore, a rate adjust factor is used to compensate for this extra demand on the satellite resources.

With the exception of the Intelnet antennas (1.2 m in diameter), all other Intelsat standard earth stations require antennas larger than our definition of a microterminal. This is also true for the private international earth stations, as indicated in their filings (Table 3.4).

The Intelnet I service is a spread-spectrum multiple-access (SSMA) service similar to the Equatorial system in the United States. It operates at 11 GHz.

The Federal Communication Commission permits unlicensed receive-only Intelnet I earth stations to be constructed within the United States. At the same time, they emphasize that unauthorized interception of the Intelnet signals is not permitted under U.S. law, as these are considered a private communications. The Intelnet II service is two-way (transmit at 14 GHz and receive at 11 GHz).

Packet Switching. Tymnet is an example of a packet-switched network for data communications. This is a mechanism for sharing communications between major nodes which may be accessed by multiple customers. Tymnet is one of several packet-switched networks in the United States. They recently instituted a microterminal connection between their nodes and selected customers. These networks connect users with data bases and computers. They are widely used for accessing information sources (such as Dialog, Newsnet, The Source, Compuserve, etc.), and their subscribers.

These terminals may also be used for remote job entry (RJE) and other

TABLE 3.3 Intelsat Earth Station Standards[a]

Standard:	A	B	C	D1	D2	E1	E2	E3	F1	F2	F3	Intelnet
Frequency bands												
Transmit (GHz)	6	6	14	6	6	14	14	14	6	6	6	14
Receive (GHz)	4	4	11	4	4	11 and 12	11 and 12	11 and 12	4	4	4	11 and 12
Typical diameter (m)	30 (15.6[b])	11	14 (11[b])	3.8	11	3.5	5.5	8	5	7	9	1.5
G/T												
Required (dBi/K)	40.7 (35.0[b])	31.7	—	22.7	31.7	25	29	34	—	—	—	18.7
Typical (dBi/K)	40.7	31.7	39 (37[b])	22.7	31.7	—	—	34	—	—	—	—
EIRP												
Global FDM/FM min. (dBW)	74.7	Same	n/a	56.6	52.7	n/a	n/a	n/a	—	—	—	—
Spot FDM/FM min. (dBW)	81.4	Same	65.7	—	—	56.7	54.5	49.3	—	—	—	—
Hemi FDM/FM (dBW)	80.0	Same	n/a	—	—	n/a	n/a	n/a	—	—	—	—
Service route	Trunk	Trunk	Trunk	Thin	Thin	IBS	IBS	IBS	IBS	IBS	IBS	Intelnet
Rate adjustment factor (IBS)	1.0	1.0	1.0	—	—	2.8	1.7	1.25	2.8	1.7	1.25	?

[a]n/a, not applicable.
[b]Revised by Intelsat in March 1986.

TABLE 3.4   Private International Earth Stations[a]

System:	PanAmSat[b]	ISI[c]	Orion[d]
Frequency bands			
Transmit (GHz)	14	14	14
Receive (GHz)	11	11 or 12	11
Typical diameter (m)	2–5.5	5–7 (Tx) 3–3.5 (Rx)	3.5–7.7
G/T, typical (dBi/K)	21.7–30.5	26.5–28	24–31
EIRP, typical (dBW)	43.4–62	60–81.2	81–84

[a]Data as of January 1987.
[b]*Simon Bolivar satellite, parameters are from the Cygnus filing (which was acquired by PanAmSat).*
[c]*International Satellites, Inc.*
[d]*Orion Satellite Corporation.*

forms of batch computer processing applications. These include payroll and corporate inventory control programs.

Arpanet includes Satnet, which is a satellite-connected (domestic and international) packet-switched network with many additional features. European ports of entry are at the Intelsat member stations in Sweden, West Germany, the United Kingdom, and Italy (see Figure 3.2).

3.3   TWO-WAY SERVICES

3.3.1   Thin Route

Telephone traffic may be categorized as either trunk (thick) or thin route services. Trunk telephony is being handled by fiber optics and is not a small business earth station application area.

Examples of thin route service are some residential, farming, and ranch services. Business communications of this type are exemplified by energy exploration (coal, oil, etc.), which may be at remote sites for extended periods, some agricultural operations, land management, and so on. Various estimates of the size of this market have been made by proponents of the Land Mobile Satellite Service (LMSS).

If the price of LMSS microterminals is sufficiently low, it may be advantageous for traditional telephone companies to use this type of satellite connection in place of the traditional copper to reach some of their more outlying customers where the cost of service and maintenance is unusually high.

In general, the LMSS services use satellite bandwidths of 2.5–5 kHz, which is too narrow for many business data services. Since neither the spectrum nor the operators have been chosen for this service, the start dates cannot be estimated nor a growth rate assigned. Eventually, very small mobile terminals will emerge, potentially in the tens of thousands for the first-generation LMSS. The growth will be spectrum limited (perhaps to 7 MHz total).

Figure 3.2. Satnet node-to-node connections in Arpanet.

The ability to use a K_u-band microterminal for voice and/or data adds to the attractiveness. If a 64-kb/s station carries a 32-kb/s voice channel, there is ample room for several medium-rate (e.g., 9.6-kb/s) data channels.

3.3.2 Emergency Services

Figure 3.3 shows the first microterminal in the United States. A devastating flood hit Johnstown, Pennsylvania, in the mid-1970s. This terminal was rushed to the scene and provided the only telephone communications out of Johnstown after the telephone building was evacuated. It provided links to the Red Cross's national headquarters in Washington, D.C., where rescue and food efforts were coordinated. The figure shows a small (1.2-m) antenna

Figure 3.3. First microterminal in the United States.

with a subreflector. The electronics (the low-noise amplifier and high-power amplifier) were mounted behind the antenna to reduce the signal loses.

This is an example of a highly transportable microterminal that can be used in the event of a natural disaster, an explosion, a nuclear meltdown, and so on. Two-way voice and data services may readily be provided.

A fire in a telephone exchange in San Juan resulted in an airline loosing its communications. Microterminals were rushed to the scene to provide seat assignment information and flight data.

3.3.3 Cellular Extensions

The cellular radio service is a land mobile terrestrial service (radio telephone) operating from a series of limited-area-coverage transmitting/receiving stations. As a vehicle moves from one area to another, the frequencies are automatically switched so that the user is not even aware of having crossed a cell boundary. Long-distance telecommunications between cellular cities may be provided either by terrestrial means (through a telephone company) or via satellite. The most likely initial business satellite application will be for billings and checking the telephone number authorization of the user. The satellite link may also be used for settling accounts between the cellular and traditional telephone companies for telephone calls placed through a cellular telephone.

4

WHY USE MICROTERMINALS?

4.1 CORPORATE CONTROL

As information management continues to grow as an element of corporate success, there will be the need for an assured supply of telecommunications at all times, in sufficient quantities at stable prices at high quality and with the flexibility to relocate terminals to reflect the inevitable changes in the network architecture.

For a telephone company these have become difficult requirements to fulfill. Each year corporate America is more demanding in its needs for digital services. Until the existing copper-based telephone networks are converted to fiber (at both the long-distance and local levels), facilities will often be stretched beyond their capabilities.

Cost containment is an important aspect. In general microterminal systems have a 4- to 5-year payback and a savings of several million dollars based on the present toll charges for terrestrial services. As the terrestrial tolls increase (mainly due to local costs) this payback period is shortened. There are intangible value-added benefits from having corporate networks that can be rapidly reconfigured to meet changing business conditions.

4.1.1 Quality Control

In the final years of the Bell System, AT&T was able to tariff a 99.5% availability for its digital services. With the divestiture of the Bell System, AT&T lost control over the Local Exchange Companies (LEC or BOCs) and the end-to-end quality control. The present digital service availability is in the range 99.7% to 99.5% (depending on the service type, LEC, exchange switch type, local line type and length, etc.). When trouble is encountered it is difficult to determine which company is at fault (or even which to complain to). The split responsibility has lead to AT&T offering to take a coordination role (for a fee) to resolve these problems, but the problems continue and the time to solve problems is not getting much shorter. Even a private leased terrestrial line encounters these problems.

One approach is to start all over with another medium and make sure that, as far as possible, all control is kept inside the user's organization. Two recent media permit this type of system: fiber optics and microterminal satellite services. The transmission medium may be leased (under a fixed or limited increase price contract) and the terminal equipment may be purchased outright. Quality control is now within the corporation's control. If a new corporate location is established to meet a requirement, the microterminal can beat the movers to the facility. No more waiting weeks or months for the telephone company. The equipment has become so simple that it does not take a genius to set it up and keep it running.

4.1.2 Privacy

As increasing amounts of data about a corporation's plans, operations, sales, and revenue flow between the decentralized corporate offices, there is more concern over privacy. A corporate network follows a limited number of nodes (thus reducing the chances of accidental interception). The use of microterminals and satellites makes the interception issue very important. In a busy digital microterminal network, intercepting an important piece of data may be difficult, but it is not impossible. Encryption at the source (telephone, PC, etc.) is the best policy. Encryption at the microterminal may be an acceptable alternative in many nongovernment cases. Combined PC/microterminals with encryption may become popular.

4.1.3 Convenience

There have been major benefits from divestiture of AT&T, but the frustrations of trying to reconfigure existing networks, and the establishment of new facilities, has been high. Figure 4.1 shows the corporate elements in a simple point-to-point network. Multiple carriers are involved.

A typical telephone call from the Communications Center involves a key-switched telephone system leased from *Bell Atlanticom*, a *Regional Bell Operating Company (RBOC)*, our own internal wiring, a 5-km overhead twisted-pair cable to the nearest local exchange carrier (*C & P of Maryland*, a *BOC*, which is part of the regional Bell Atlanticom family), a 10-km intra-LATA T1 supplied by *C & P of Maryland* to the main exchange for some of the lines, a 15-km inter-LATA T1 carrier connection supplied by *AT&T Communications* to our main exchange, where the call may be routed by another

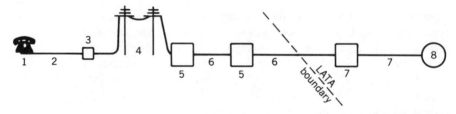

Element	Sources
1. Telephone	AT&T
	RBOC
	Department stores
	Specialty shops
	Mail order
2. Customer premises wiring	BOC
	Electrical contractor
	Do-it-yourself
3. Protective device	BOC or purchase
4. Local loop	BOC
5. Local exchanges	BOC (LEC)
6. Interexchange:	
Intra LATA	BOC
Inter LATA	AT&T
7. Long-distance exchange and lines	AT&T
	U.S. Sprint
	MCI
	etc.
8. Other end (inverse of 1 to 6)	

Figure 4.1. Split responsibility.

T1 carrier to our long-distance carrier's (*U.S. Sprint*) central office, where any combination of owned or leased lines (from AT&T, a BOC, or another common carrier) is used to connect to Sprint's fiber and satellite transcontinental network.

If the call is to California, the distant end may involve *Pacific Telesis* (an *RBOC*), *Pacific Bell* (an *LEC*), *General Telephone of California* (an independent telephone company), *AT&T Communications,* and so on.

In case of trouble, it is difficult to find a single point of responsibility. A microterminal network circumvents these problems by a direct connection (via the hub) between corporate elements.

Convenience also involves the ability to restructure a corporate network as older facilities are closed and new ones opened. A microterminal can be removed and set up at a new location within a day (substantially shorter than the telephone company response time). Temporary networks can be established for:

• Proposal teams scattered across the nation

- Retooling for a new production year with computer-aided design (CAD) and computer-aided manufacturing (CAM) facilities linked directly to the manufacturing facilities
- Field offices for conducting performance tests, and so on.

After the mission has been fulfilled, the terminals may be moved to a new location, or storage, until the next requirement arises.

Convenience is also manifested in the system transparency. Most data users will have a hard time telling the difference between the microterminal and traditional services. Eventually, the improved bit error performance and a slightly increased throughput will be evident. As higher-speed services are introduced, the difference becomes more apparent. The response time is the same or faster than that of a normal packet-switched network.

In one retail store credit card verification network the response time dropped from 40 seconds (telephone) to 6 seconds (by microterminal). When tens of thousands of transactions are involved the cumulative labor cost savings are impressive.

4.1.4 Independent Facilities

Sometimes it is nice to have a redundant path for at least some telecommunications to enable business to be continued during natural disasters, accidents, telephone strikes, and so on. By using an on-premises microterminal, telecommunications will continue during these periods.

4.1.5 Reliability

There are three basic reliability elements in a microterminal network: local connections, network equipment, and weather.

Local Connections. These are the in-plant connections between the end device (e.g., a word processor terminal) and the microterminal. They are subject to all of the same problems encountered in a normal telephone system. The run length is shorter, entirely on the user's premises, and usually does not need switching; therefore, the opportunity for problems is smaller. In some situations, the local PBX (private branch exchange) will be connected to the microterminal. The T1 modem (if used) may be located within the PBX, thus simplifying the PBX/microterminal connections. If X.25 packet switching is used, the packet assembler/disassembler may be integrated into the microterminal controller.

Microterminal Network Equipment. This element consists of the local microterminal controller and transmit equipment, the hub equipment, and the distant microterminal reception equipment and its controller. Since low power levels and moderate bit rates are involved, this equipment can be designed for a long mean time between failures (MTBF).

In most microterminal applications redundancy has been eliminated (to cut the cost based on the trouble-free history of these stations). If a customer wants a redundant station, there are three methods: (1) overbuild the existing station with two of everything and add switches and a control system with a cost of 2½ times the nonredundant station; or (2) install two nonredundant stations at the critical locations. At the low per-station prices, the second may be the best choice. When both stations are operable, twice as much traffic capacity is available for peak hours, crisis, and so on. When one station is not working, it is replaced by the spare. The mean time to replace (MTTR) a failed element is 24 hours or less using courier delivery of replacement units; or (3) to have a spare set of the critical units at each earth station (costs about 1¾ times as much as a nonredundant station).

We favor the second approach as being more cost and traffic effective for those situations where the equipment reliability must exceed the 99.9–99.99% (52 minutes to 8.75 hours of outage per year) of the nonredundant station.

The hub presents a special problem. If the hub fails, the entire network will not operate. It should be internally redundant but not necessarily to a 100% level. In very critical networks, a second hub should be considered. The second hub could be sharing the load with the primary hub or it may be in a hot-standby mode (ready to assume control on a moment's notice). The second hub could be located at the hub of another system. If both systems agree to serve as mutual backups, there will be two pools of facilities and operators.

Weather. Weather is a problem in both the uplinks and downlinks at K_u-band. There are two ways to overcome this natural effect. The first is to increase the uplink power during a rainstorm.

By monitoring the 12-GHz satellite beacon, a hub station can sense its local rain attenuation and boost its uplink power by an appropriate amount. If it can see one of its own carriers (translated to 12 GHz), automatic level control (ALC) can be used to maintain service until the rain attenuation exceeds the station's capabilities. Considering the low price of the microterminals, uplink power control is not considered to be economically feasible except at the hub.

The second method is to use a redundant path. Two hubs in climatically diverse areas would be needed. The first of these was established by Williams Pipe Line with hubs in Iowa and Oklahoma. The main station is located in Tulsa. A fiber optics link connects the two hubs. Both can function as transmitters and receivers. The actual controllers are located in Tulsa. The alternate location (Des Moines) is a slave to Tulsa in case of bad weather, transmitter/receiver problems, maintenance, and so on (see Figure 4.2). In addition, the Communications Center has planned a series of tests of satellite diversity to see if it is possible to overcome local rain conditions from a single site.

In a data application, rain attenuation may cause an outage of a few sec-

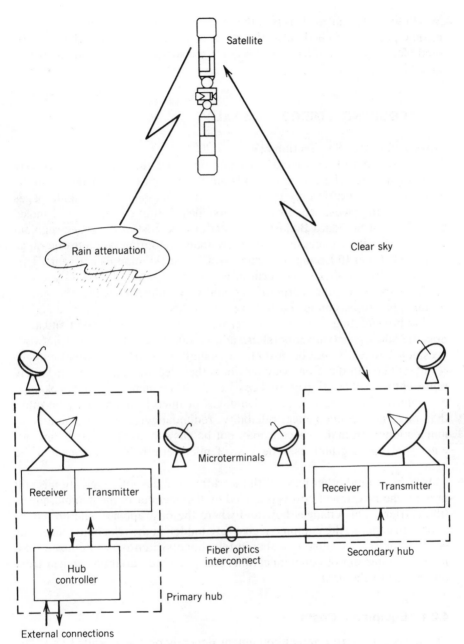

Figure 4.2. Diversity hub configuration.

onds to several minutes. This is in the range of the delays that are encountered in some packet-switched networks at the peak busy hours and thus may be tolerable. Voice and video demand real-time services, so any outage is less tolerable.

4.2 ACQUIRING A MICROTERMINAL

Cross Reference: See Section 6.3.

For a number of customers, microterminals are the least expensive way to communicate. The choice depends on the amount of traffic placed in the network. If the traffic only consists of a few messages of facsimile pages (e.g., three) the investment cost will be too high. If traffic is high (i.e., handled by an Intelsat Standard B class earth station) the cost of the space segment and the hub will be excessive. Microterminals are optimal for a traffic range between 0.1 and 10 Erlangs (1 Erlang = 83,077 paid minutes per year). This level of traffic implies some form of a local area network (LAN) within a store, shopping center, corporate complex or industrial park. The micro-terminal has to be connected to the user's LAN.

Hub broadcasting to many microterminals represents a different situation. Specific addresses (subscribers) are addressed. When this type of network is established it may not be cost effective until there are a sufficient number of subscribers. In these one-way networks the telecommunications costs may represent a small fraction of total cost of acquiring, managing, and packaging raw information. The primary advantages of microterminals may be other than cost (i.e., convenience, reliability, redundancy, and true digital com-munications). In many cases it may not be easy to compare costs and ad-equacy of the alternatives. The cost of microterminals may appear to be high.

In Europe (and other parts of the world) where despotic regulation is still present, the investment cost is provided by the operator of the microterminals (the carrier). This changes how and where the cost appears and on whose books. In general, the equipment costs are hidden in the telecommunications charges for service, just as a shared telephone switch is not itemized on a monthly statement of communications charges. Odd variations appear from one nation to the next.

4.2.1 Equipment Costs

The costs for microterminal equipment depends on the intended use. The C-200 Micro Earth Station from Equatorial Communications costs about $6000. This terminal operates at C-band and provides 9.6-kb/s two-way ser-vice. There is a large base of technology supporting C-band equipment which tends to bring down the price of an earth station. Another factor in the low price is the large quantity of production by a single vendor, allowing the full impact of "economy of scale." The $6000 price is for quantities of 500 terminals or more. The price of the earth terminals depends to some extent on the data rate that the terminal is required to handle, with higher data rates requiring more expensive modems, and so on.

4.2.2 Purchased Systems

The various vendors of microterminal systems have different ways of pricing their systems. Most of them offer the option for the customer to purchase the microterminals, and in some cases also the hub station. In other cases, the hub station is furnished by the vendor, and its costs are shared by a number of users. The user still has the option to purchase its own micro terminal. There is a monthly fee for the space segment service. Hub station costs depend highly on the system design and whether the hub station serves only one customer or is shared among a number of customers. Table 4.1 estimates the hub costs.

4.2.3 Leased (with an Option to Buy) Systems

There are a great variety of products for the home and office that can be leased with an option to buy at a later time. Among these products are automobiles, computers, copiers, PBX telephone systems, and so on. Even if

TABLE 4.1 Hub Station Cost Elements[a]

	Initial Cost	Cost per Month	Notes
Equipment Costs			
Hub station equipment	$850,000	$ 17,645	9% interest, 5 yr
Hub station land, roads, etc.	275,000	3,484	9% interest, 10 yr

	Annual Cost	Cost per Month	Notes
Operating expenses			
Personnel (4 @ $35,000/yr)	140,000	11,667	(one per shift)
Billing personnel			
(12 @ $17,500/yr)	210,000	17,500	
Overhead (including	525,000	43,750	(150%)
maintenance)			
Subtotal		94,046	
General and administrative		4,702	at 5%
Profit		9,875	at 10%
Total		$108,623	per month

Networks sharing the hub:			
1	2	3	4
Cost per month[b] $108,623	$54,312	$36,208	$27,156

[a]Includes the antennas, RF control, billing, and an initial set of baseband port connections. As the system grows, additional port costs will be encountered. As it gets more sophisticated and additional protocols are added, ancillary equipment costs will grow. The space segment is not included.
[b]n-way split of $108,623. Limited additional equipment expenses will be incurred for network-unique equipment.

the equipment is purchased outright, a corporation is likely to sell the equipment to a bank or credit corporation and lease back the microterminals. This frees up the cash for reinvestment. Table 4.2 shows the potential lease costs.

4.2.4   Rented Systems

A number of vendors offer microterminal services on a rental basis. This arrangement is favorable for those customers that are not interested in the maintenance, service, installation, or modification of the equipment. It is also a reasonable arrangement for the vendor who can arrange for installation and maintenance on the basis of his overall needs, which may include several networks. Rentals are attractive to companies that need only a temporary network (seasonal, construction project, etc.) or have a limited credit capability.

4.2.5   Total Costs

Tables 4.3 and 4.4 provide cost estimates for typical K_u-band and C-band services respectively.

TABLE 4.2   Microterminal Leases[a]

Line				
1	Cost basis of the terminal	$25,000	$15,000	$6,000
2	Useful lifetime (yr)	10	10	10
3	Basic lease term (yr)	7.5	7.5	5
4	Residual value at end of lease (10% of line 1)	$ 2,500	$ 1,500	$ 600
5	Amount financed (line 1 − line 4)	$22,500	$13,500	$5,400
	Prime rate 8.5%			
	Risk 1.0%			
6	Interest 9.5%			
7	Period (months)	90	90	60
8	Monthly payment	$ 350	$ 210	$ 113
9	Maintenance (0.008 × line 1) per month	$ 200	$ 120	$ 48
10	Total monthly (lines 8 + line 9)	$ 550	$ 330	$ 161
11	Buyout at end of the lease (line 4)	$ 2,500	$ 1,500	$ 600
12	Total cost during lease (line 7 × line 10)	$49,545	$29,727	$9,685
13	Cost per month after buy out[b]	$ 200	$ 120	$ 48

[a]Space segment is not included, some lines have been rounded.
[b]At this point, the station has been paid for and fully depreciated. The recurring expense is the maintenance (line 9).

TABLE 4.3 Tridom's Cost Estimate

Remote communications stations	$8000–$10,000 each
Monthly maintenance	$60–$100 each
Remote station monthly space segment cost	$10–$200 each
Hub	(not specified)

Source: L. Huang, "Small-Aperture Earth Stations Can Be an Alternative to Private Line Networks," *Communications News,* March 1986.

4.3 FUTURE EXPANSIONS

The versatility of these little terminals should not be underestimated.

In the case of microterminals, it is easy to expand the network by adding additional stations and occasionally adjusting the space segment. A well-conceived hub can readily be expanded as the network grows. The major changes will come when the type of traffic is altered (from 64 kb/s to T1 or to video).

Eventually, the network may be slowly reconfigured to reduce the space segment requirements if these begin to get into short supply. The retrofits will involve the use of less resources through more efficient encoding (fewer bits per information element) and modulation (more bits per Hertz). These are discussed in Chapters 13 and 14. The small terminal antennas may be enlarged if the trade between larger antennas and reduced space segment costs is favorable. As the price of microterminals continues to fall, and the space segment charges increase, the original trade-offs may need to be revisited.

TABLE 4.4. Equatorial Costs for a 500-Station Network, 1986

Operational	
Space segment (at $17,500/month)	$ 35/month
Network manager (hub)	35/month
Local maintenance	60/month
Per station	$130/month
Original (One-Time)	
C200 two-way terminal	$6200
Installation	1200
Miscellaneous	30
Total	$7430

4.4 ALTERNATIVE SOURCES OF SATELLITE SERVICE

The earth segment is available from many vendors and full-service suppliers (see, e.g., Chapter 7). The space segment needs can be fulfilled by a variety of sources (see Table 4.5).

4.4.1 Traditional C-Band Networks

As shown in Table 4.5, there are a variety of satellite capacity sources. As the amount of unused and unreserved capacity shrinks, the price may increase. Satellite time is available by the hour, month, year, and spacecraft lifetime. Capacity is available on a fully protected, unprotected, and preemptible basis (see Chapter 7).

The C-band networks share their frequencies with terrestrial point-to-point

TABLE 4.5   C-Band Space Segment Suppliers

C-Band capacity
Uplink: 5.925–6.425 GHz
Downlink: 3.700–4.200 GHz

Source	Satellite(s)[a]
Alascom, Inc.	*Aurora I*[b]
Contel-ASC	*ASC-1*
	Westar[c]
	Galaxy III[cd]
	[ASC-2]
AT&T Communications	*Telstar 301*
	Telstar 302
	Telstar 303[d]
Comsat General	*Comstar D4*
GTE Spacenet	*Spacenet I*
	Spacenet II
	[Spacenet III]
Hughes Communications	*Galaxy I*[d]
	Galaxy II
	Galaxy III[d]
GE Americom	*Satcom IR*
	Satcom IIR
	Satcom IIIR[d]
	Satcom IV[d]
Western UnionTelegraph Co.	*Westar III*
	Westar IV
	Westar V
	[Westar VI-S]

[a][ ], satellite awaiting launch as of mid 1987.
[b]Too far west for good east coast service.
[c]Portion
[d]Primarily a video satellite.

microwave networks. This limits where conventional earth stations may be located. The traditional C-band SCPC services used a large antenna to reduce terrestrial interference levels.

Equatorial has been able to use small (60-cm) receive-only and 60 cm × 120 cm transmit/receive antennas, through the use of spread-spectrum multiple access, which tolerates a lot of interference while producing little to other systems (see Chapter 5).

4.4.2 K$_u$-Band Networks

Table 4.6 lists the U.S. K$_u$-band satellites.

The K$_u$-band satellites have higher power and terrestrial interference-free frequencies. These attributes make them more desirable for many micro-terminal uses.

4.5 FULLY DIGITAL NETWORKS

The Communications Center held a conference on fiber optics in 1986 for its clients and other attendees. There was a lively discussion among the 100 attendees as to when the Integrated Switched Digital Network (ISDN) would arrive. The general feeling is that it is coming, but slowly. In 1987, only 20% of the U.S. traffic is digital (either on voice or digital grade lines), but in

TABLE 4.6 K$_u$-Band Space Segment Suppliers

K$_u$-band capacity
Uplink: 14.0–14.5 GHz
Downlink: 11.7–12.2 GHz

Source	Satellite(s)[a]
Contel-ASC	ASC-1
	[ASC-2]
Comsat General	SBS-1
GTE Spacenet	Gstar I
	Gstar II
	[Gstar III]
	Spacenet I
	Spacenet II
	[Spacenet III]
MCI	SBS-2
	SBS-3
IBM Satellite	SBS-4
	[SBS-5]
GE Americom	Satcom K-1[b]
	Satcom K-2[b]

[a] [ ], satellite awaiting launch as of mid 1987.
[b] Primarily a video satellite.

some cities 50% of the business district telephone lines carry data via modems during the day.

The more important aspects that emerged was the rapid growth of digital services and the feeling among some of the corporate members of the audience that bypassing the local exchange carriers was needed to fully attain a digital service. The long-haul telecommunications networks are rapidly converting to fiber which is all digital. The local loops to these fibers is the source of the high bit error rates and other service problems. The microterminal is the ultimate bypass method that has been selected to provide bit error rates in the range 10^{-6} to 10^{-9} (one error in a million or billion bits).

5

WHO ARE THE USERS?

5.1 MICROTERMINAL USERS

5.1.1 Means of Serving Users

There are several basic methods of supplying limited bit-rate services to customers with microterminals. One method uses single channel per carrier (SCPC) for the small station to hub, and time-division multiplex (TDM) for the hub to small link.

The second method employs spread-spectrum multiple access (SSMA). Equatorial Communications is also in the SSMA communications service business, furnishing low-speed data transmissions to thousands of receive-only customers (Figure 5.1), and providing a two-way data service (Figure 5.2) to a large customer base.

5.1.2 Industrial User Groups

The industrial user groups range over a wide variety of corporations: Schlumberger, a Texas-based oil exploration company, is using a satellite communications network to connect drilling sites to a central computer for the evaluation of drilling operations.

The outstations are transportable stations mounted in specially outfitted refuse trucks. These trucks are among the most rugged that can be easily procured for use on highways and can be operated in the hostile environment where many of the drilling sites are located.

Novanet is operating a microterminal network for HNG-Internorth, Tenngasco, and other pipeline companies. This service uses C-band microter-

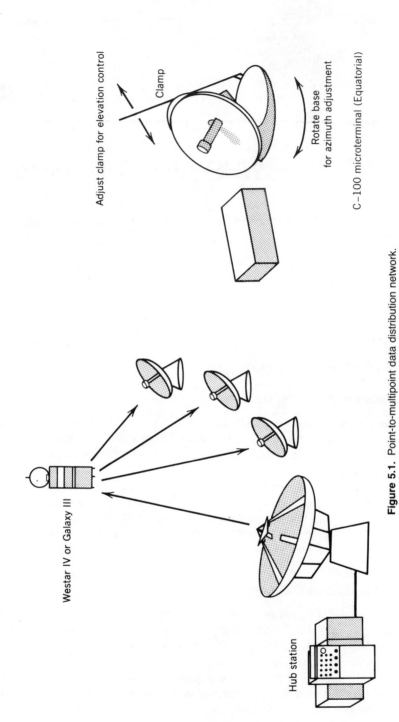

Figure 5.1. Point-to-multipoint data distribution network.

Adjust clamp for elevation control

Clamp

Rotate base
for azimuth adjustment

C–100 microterminal (Equatorial)

Westar IV or Galaxy III

Hub station

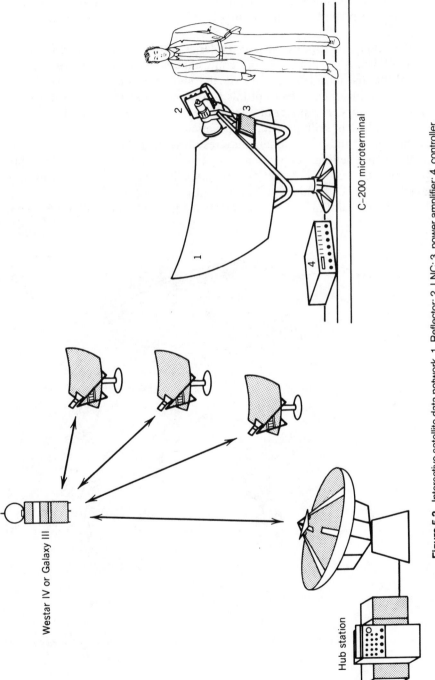

Figure 5.2. Interactive satellite data network. 1, Reflector; 2, LNC; 3, power amplifier; 4, controller.

C-200 microterminal

Westar IV or Galaxy III

Hub station

minals and replaced leased-line telephone service. The purpose of this network was to provide remote control and automated monitoring of the pipelines.

Other industrial groups include the large automotive manufacturers like GM and Ford. An internal network could expand corporate communications beyond the district offices to most dealer showrooms and major suppliers. One of the primary capabilities of this type of network is the ability to closely control inventory, raw materials, and spare parts. If the network detects that one body style is selling faster than anticipated, increased orders for the assembly lines and the suppliers may result. If the network establishes that one line is not selling, production and parts orders could be curtailed, thus avoiding the large inventories that plague all manufacturers. Just-in-time manufacturing needs this type of rapid information flow.

Hewlett-Packard (a manufacturer of test equipment, computers, etc.) has

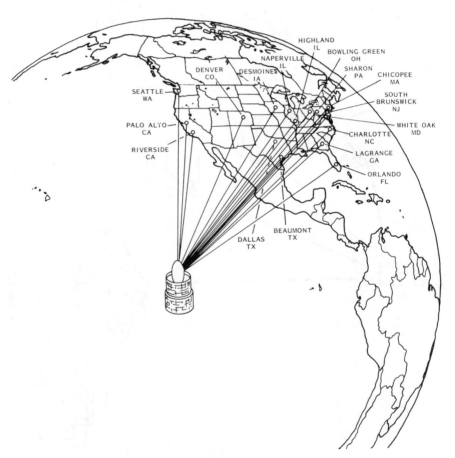

Figure 5.3. Dow Jones satellite network.

used small K_u-band terminals for business video and has planned a micro-terminal network. Newspapers and news magazines have found that remote printing (via satellite) is profitable. Figure 5.3 shows the Dow Jones network (under 100 kb/s) used to make plates to publish the *Wall Street Journal*. At the present, large C-band terminals are used.

5.1.3 Retail Networks

Wal-Mart, an Arkansas-based retail store operator, contracted for a satellite communications network to connect their retail outlets in a 20-state area with data and video services. Southland Corporation selected a TDMA satellite network to serve their 7-Eleven retail food stores with data services.

5.1.4 Document Express Services

Federal Express Corporation, a Tennessee-based document and small package delivery operator, had contracted for a very large network consisting of telephone lines, facsimile machines, and (eventually) microterminals. This network was to facilitate the transmission of facsimile pages between customer locations ("Zapmail"), but due to persistent difficulties with the temporary terrestrial telephone lines and system problems, they withdrew the plan in 1986. The terrestrial performance was "subpar to Federal Express' standards." However, we are told the microterminals worked fine (but these were not enough to carry the entire national service). It may be redesigned and offered at a later date (see Figure 5.4).

5.1.5 Financial Services

The Beneficial Corporation uses network services for their loan business. They announced a plan to buy up to 250 microterminals with options for extending the networks. Citishare (a part of Citicorp/Citibank) has acquired terminals for their transaction network at several hundred sites. Among the more active microterminal users is Merrill Lynch, one of the leading North American stockbrokers. The Farmers Insurance Group (see Figure 5.5) may expand beyond their original 1000 terminals.

Xerox Data Services provides time-shared computer services (inventory control, sales records, payroll, etc.) via microterminals and its computer system in Hawthorne, California (see Figure 5.6).

One of the earliest users of microterminals were the commodity networks. The one-way service provided data on the sale prices for grains, pork, and so on.

Another type of business that might be attracted to microterminals is one that is seasonal or moving. An example of this might be the H & R Block tax preparation service. In the United States, income taxes are due for filing with the Internal Revenue Service on or before April 15. H & R Block is a

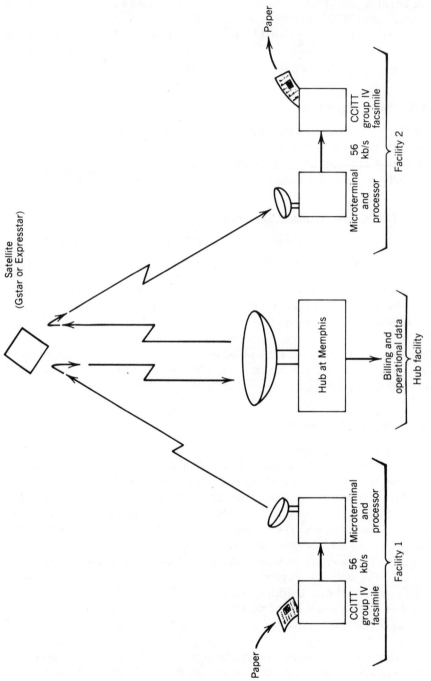

Figure 5.4. Federal Express's original plan.

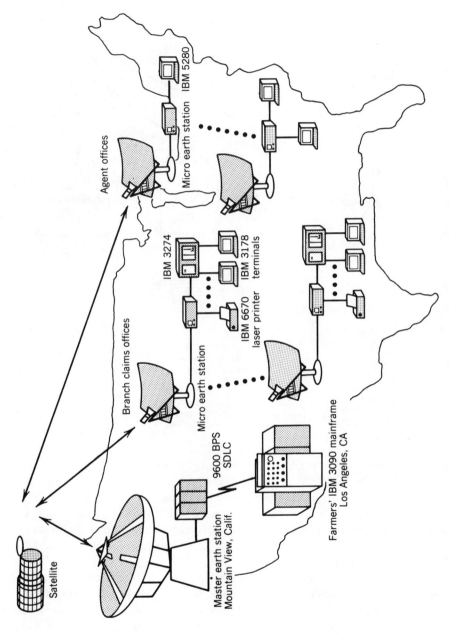

Figure 5.5. Farmers Insurance Group network.

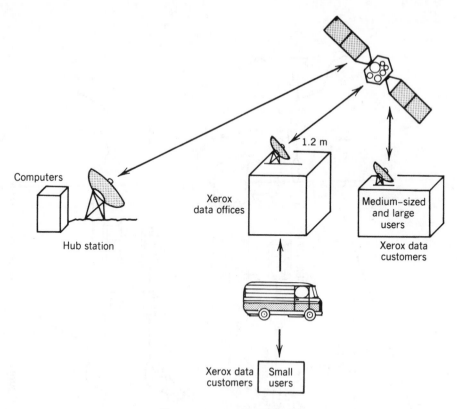

Figure 5.6. Xerox's network.

large tax preparation service. They rent empty stores for the busy period (January through April 15). After the taxes are filed, the stores are closed. When the next tax year comes, the store may be unavailable and, therefore, new quarters must be found. This results in a highly mobile business which could make greater use of a two-way digital telecommunications link to their master computer for the computation of taxes and to reference a tax law library.

Remote automatic teller machines are another potential market. In the case of some banks there is a requirment for two independent telephone lines (sometimes through two separate telephone switching centers) between the ATM and the bank. This redundancy is done to improve reliability and to eliminate the potential loss of the ATM in case of, for example, a fire in a telephone switching center. At present, ATM machines must transact at least $3000 per month to pay for their cost, maintenance, and restocking. Many of these machines are now being located in grocery stores and other points of purchases. Regionwide (and eventually nationwide) ATM banking networks are emerging in the United States. This represents another potential opportunity for microterminals.

5.1.6 Weather Reports

In addition to the National Weather Service (provided by the government), there are many specialized private weather services providing fine-grain interpretive services to highway departments, farmers, ski resort operators, pilots, and radio/TV stations. Accuweather, for example, uses microterminals to disseminate digital weather maps to TV stations and other clients.

5.1.7 International Services

Equatorial has indicated an interest in extending its service outside the United States by connecting to Intelsat's Intelnet I (one-way) and Intelnet II (two-way) international speed-spectrum satellite services. An example is the direct reception of the Reuter's news service from the United Kingdom. In addition, Equatorial is expanding to Australia (at K_u-band), India and South America.

5.1.8 Summary

Table 5.1 shows the status of the various networks and their suppliers and satellites.

5.2 TRAFFIC DISTRIBUTION

5.2.1 Where Are the Microterminals Located?

The Communications Center acquired a data base listing all licensed stations. We identified those stations under 2.8 m. A detailed study prepared by the Communications Center for a city municipal client in the southwestern United States showed a close correlation between the location of satellite microterminals and the areas of the most rapid business growth.

Many of the microterminals are located in small towns, poorly served by traditional means.

5.2.2 How Does the Traffic Vary?

Data originated in an office environment (word processing, access to data banks, etc.) follow a normal business telephone traffic pattern (see Figure 5.7). Retail data have several peaks. Credit checks are done during store hours with weekday peaks at noon, 3 to 8 P.M. and weekends (9 A.M. to 9 P.M.). Inventory control, sales reports, and other store data may enter the system at anytime, but particularly in the morning (8 to 10 A.M.) and after the store has closed (9 P.M. to 8 A.M.). Nonurgent data (total sales, etc.) can be stored and later polled by the hub during times when the system would otherwise by idle. Gasoline sales tend to follow the retail sales, except that the peaks are flattened and some transactions take place after 8 P.M. and many before 9 A.M.

TABLE 5.1   Microterminal Networks

User	Supplier[a]	Start	Locations in Mid-1986 (Future)	Use	Satellite or Band
Merrill Lynch	Eq 2	1985	1000	Financial	*Westar IV*
J.C. Penney, New York	PSN	1982	150	Sales and training	*Gstar*
Xerox Data Services, Hawthorne, Calif.	Telecom General	1985	100 (800)	Batch processing	*Gstar*
Farmers Insurance Group, California	Eq 2	1985	1500 (3000)	Data base access	*Galaxy III*
Citibank, New York	Eq 2 Telecom General	c.1984 1986	175 500	Stock transactions	*Westar IV* *Gstar*
Niagara Mohawk, upstate New York	Eq 2	1985	(900)	Power lines	*Galaxy III* *Westar IV*
Schlumberger, Houston, Tex.	M/A-Com	1984	500	Oil	*SBS-3*
Southland Corporation, Dallas, Tex.	M/A-Com	1986	(300)	Retail credit	*Gstar*
Wal-Mart, Bentonville, Ark.	M/A-Com	1986	(800)	Retail sales	K
K-mart,[b] Troy, Mich.	(GTE)	1987	(2100)	Retail credit and video	GTE-K
U.S. National Weather Service, Suitland, Md.	Eq 1	1984	1000	Weather	C
N.Y. Stock Exchange, New York	Eq 1	1985	(4000)	Stock data	C
Federal Express, Memphis, Tenn.	M/A-Com	1986	100	Internal Communications	*Gstar III*
Commodity News Service	Eq 1	1980	100	Commodities	*Westar IV*
Reuters Limited, New York	Eq 1		?	News	*Westar IV*
Associated Press, (AP) New York	Telecom General and Eq 1	1985	200 ?	News	C and K
United Press International (UPI) New York	Eq 1	?	150 ?	News	C

TABLE 5.1   *Continued*

User	Supplier[a]	Start	Locations in Mid-1986 (Future)	Use	Satellite or Band
Market Information, United Telecom	Eq 1	?	?	Stocks	C
Tymstar, McDonnel Douglas St. Louis, Mo.	(self)	1985	?	Packet	K
Halliburton	COMSAT	1986	?	Oil	K
Beneficial Corp.	Eq 2	1986	?	Financial	C
Massachusetts Indemnity & Life	Eq 2	1980	1000	Insurance access	C
HNG Internorth (gas pipeline co.)	Eq 2	1986	?	Oil pipeline	C
Tenngasco	Eq 2	1986	?	Oil pipeline	C
Commodity Quotations	Eq 1	1984	500	Commodity	C
Lotus Cambridge, MA	Eq 1	1986	(?)	Stockmarket	C
Bridge Market Data	Eq 2	1986	(25)	Stockmarket	C
Williams Pipelines Tulsa, Ok.	Comsat	1986	(120)	Control	K
A.L. Williams Insurance	Sci. Atl.	1986	(450)	Video–4.5m	*Gstar I*
Service merchandise	M/A-Com	1986	(300)	Automation and materials handling	K
Rural America Comm.	Adcom	1986	(120–15,000)	Data processing	*Satcom K2*
Small Terminal Video					
Ford Motor Company Dearborne, MI	Sci. Atl.	1986	(?)	Training and video	K
Private Satellite Network New York	G.I.	1982	1000	Business TV	*Gstar*
Merrill Lynch New York	PSN	1982	—	Business TV	*Gstar*

[a]Eq 1 and 2 denote the one- and two-way Equatorial services. Comsat's microterminal facilities have been acquired by Contel-ASC and M/A-Com's operation is now part of Hughes. G.I. is General Instrument and Sci. Atl. is Scientific-Atlanta.
[b]K-mart includes Kresge Variety Stores, K-mart, Waldenbooks, Jupiter Discount Stores, Furr's Cafeterias, Bishop's Cafeterias, Designer Depot Stores, and Pay Less Stores.

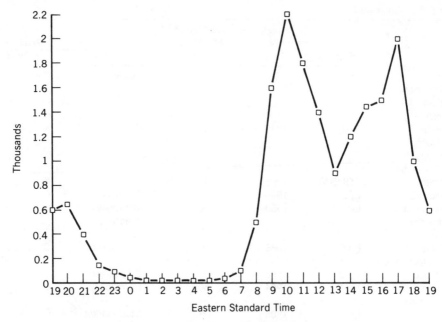

Figure 5.7. Telephone traffic (business day).

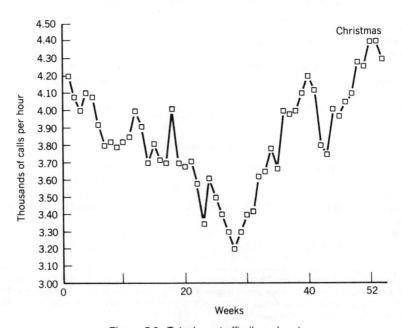

Figure 5.8. Telephone traffic (busy hour).

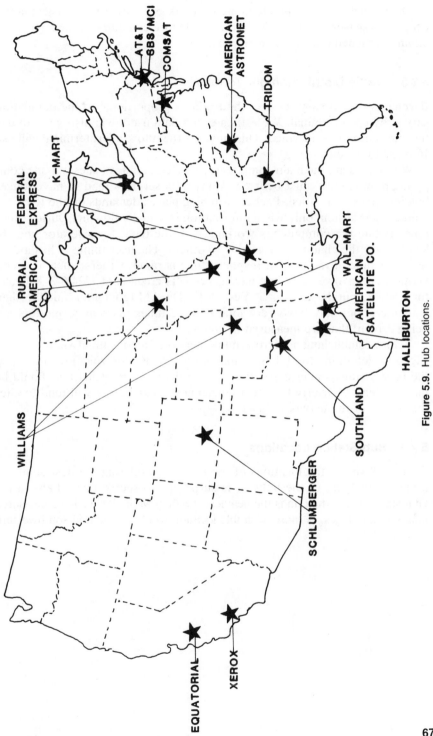

Figure 5.9. Hub locations.

The annual variation in telephone traffic (see Figure 5.8) is busiest in the Christmas season, particularly in the retail store communications. The minimum is reached during summer vacations.

5.2.3 Traffic Distribution per Terminal

If voice is the primary service, the traffic is proportional to the population served by each terminal. We estimate that data users will also want to use the microterminals for voice. This suggests the future microterminal will use 16 or 32 kb/s voice.

Where data are the main component, a different index (sales, on-site computing power, etc.) must be used. Many data services tend to have longer holding times than voice. Packet switching places demands on the telecommunications system only when information is actually flowing. During other times (while the computer is working with the data, the operator goes to lunch, etc.) the channel is available to others. The user thinks it has a dedicated channel if the system response time is rapid. Others, such as credit verification, involve very little intelligence ("Is credit card 1234-678-900 valid for a $45.45 purchase?" . . . "Yes."). CAD/CAM (computer-aided design/manufacturing) may have very long holding times (as remote plotters are being driven) that are measured in hours.

Before establishing a microterminal network, a consultant should be engaged to determine the kilobit minutes (or Erlangs for voice) of traffic, so the proper transponder capacity can be ordered. A reevaluation should be made at least quarterly to detect changing use habits that will tend to alter the profile of the network's users.

5.2.4 Hub Station Locations

Figure 5.9 shows the locations of the hubs. Considering the desirability of a high availability of the hubs, it is surprising to see so many hubs in the Gulf states where the rain is the heaviest. Even in Memphis we have measured uplink rain outages. A more desirable location would be the arid southwestern United States.

6

HOW IS THE SERVICE PROVIDED?

6.1 ACCESS METHODS

The space segment has three resources: time, bandwidth, and space (polarizations and beams). Figure 6.1 represents these three assets as axes of a cube.

6.1.1 Single Channel per Carrier

Single channel per carrier (SCPC) access to the satellite is probably the simplest method of accessing a satellite system. In a microterminal network, SCPC is used primarily from the small terminal to the hub station. Access in a microterminal network can be organized in essentially three ways:

1. When the number of microterminals is small enough and the available satellite bandwidth is large enough, each small earth station can have its own unique channel assignment.
2. Same as above, but cofrequency stations time share the channel under supervision of the hub.
3. When the number of small stations is large or the amount of available bandwidth in the satellite is limited, it may be necessary to change frequency assignments continually to find available channels from a pool.

An example of an SCPC system is a network in which the outlying terminals are assigned fixed-frequency assignments for transmission to the hub

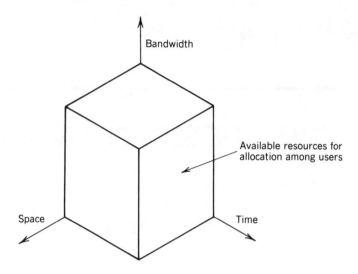

Figure 6.1. Resources.

station. In one system the channel rate is 56 kb/s and modulation is BPSK with rate two-thirds encoding. This yields an information rate of 37.333 kb/s. This access system allows the outlying terminal electronics to be relatively simple. Each terminal is assigned a frequency that is controlled by a crystal oscillator. The stability and accuracy of the crystal oscillator are determined by the channel spacing requirements, which in turn is a function of the network organization and the transmitted data rate.

For networks with QPSK modulation, the channel bandwidth can be as little as 45 kHz (or less), so it is important to have accurate crystal oscillators. For example, the Intelsat SCPC transponders operating at 6 GHz require a frequency accuracy of ±250 Hz to be maintained at each earth station. This frequency accuracy requires crystal oscillators with stabilities of $\pm 0.5 \times 10^{-7}$. To obtain accuracy of that order of magnitude requires ovenized crystal oscillators, particularly in view of the fact that these devices must operate in the outdoor unit, which is exposed to the extremes of temperature.

To obtain a large number of different frequencies for the outlying stations, one requires either many different crystals for the oscillators or a frequency synthesizer which is driven by a fixed crystal oscillator generally at a frequency which is a multiple or submultiple of the channel spacing. Employing synthesizers to generate the required transmit frequencies allows the crystal oscillator to be the same for all of the outlying earth stations, thus simplifying the maintenance logistics.

The second method of sharing SCPC channels can also be implemented using a synthesizer which is driven by a signal received from the hub station. This signal indicates which channels are available for the outlying terminal to use for its transmission. That channel can then be remotely selected by electronically changing the synthesizer to produce the required output transmit frequency.

Most networks that use SCPC for the transmission from the outlying station to the hub station usually do not use SCPC in the reverse direction. Some form of time-division multiplex is often used on a single wideband carrier from the hub station to the outlying station. The outlying station then picks out its own receive signal at the proper time from the TDM sequence. This direction usually runs at a much higher speed than the SCPC, for example, at a 16.8-Mb/s data rate.

The early network architectures vary considerably from vendor to vendor and from user to user, with each network organized to satisfy a requirement imposed by the vendor or the users and the data rate requirement in each direction.

6.1.2 Time-Division Multiple Access

Time-division multiple access (TDMA) is an access scheme used to optimize the satellite transponder resources (see Figure 6.2). In time-division multiple access, several stations sequentially access the transponder for short periods of time. There is only one carrier in the transponder, and therefore each earth station could saturate the transponder to provide the optimum output. The outlying terminals in a microterminal network normally cannot saturate the satellite transponder, so the advantages in using single carrier per transponder time-division multiple access for the small terminal to the hub station

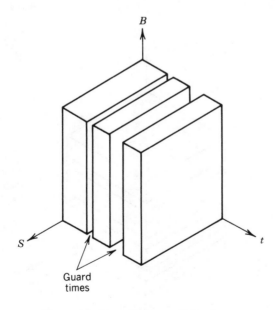

Figure 6.2. Time-division multiple access.

in such a network may be limited to the elimination of the higher-stability local oscillators needed in the narrow-channel SCPC or TDMA service. In a multichannel transponder, TDMA may be used in each channel. Many stations may be assigned to time-share each channel, but only one is permitted to transmit at a time. The time-division multiple-access system requires that the small earth station keep time and send its transmissions at the appropriate time slot.

On the return link from the hub to the small stations, it is possible to use time-division multiplex (TDM), in which case, a single station multiplexes many messages to the various outlying stations in time sequence. This scheme does not have the same complexity as time-division multiple access since the timing and time sequencing is performed only at the hub, and therefore is under much better control than if timing had to be synchronized among a number of different stations.

6.1.3 Frequency-Division Multiple Access

Frequency-division multiple access (FDMA) is another form of accessing the satellite resources. The difference between frequency-division multiple access and single channel per carrier access is that while the transponder is split into many subbonds in FDMA, any of the carriers may have multiple channels (using subcarriers) using frequency-division *multiplex* (FDM). Thus FDMA may (but does not have to) be a multiple channel per carrier (MCPC), whereas SCPC can have only one channel and multiple SCPCs may share a transponder. This is a form of FDMA (see Figure 6.3).

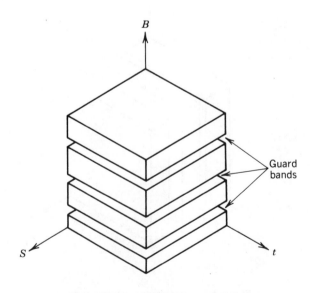

Figure 6.3. Frequency-division multiple access.

6.1.4 Spread-Spectrum Multiple Access

Spread-spectrum multiple access (SSMA), which is sometimes also referred to as code-division multiple access (CDMA), has been used for military satellite communications. One commercial operator has chosen to use spread-spectrum access to the satellite for reasons of spreading the transmitting power rather than for security.

Equatorial Communications Company uses spread-spectrum multiple access with a spreading width of 5 MHz to transmit data rates of up to 19.2 kb/s. By assigning different codes to a pseudorandom spreading series, it is possible to obtain noninterfering code-division multiple access to the same spectrum in a satellite. The power required for each carrier has to be sized such that it is sufficient to provide the required bit error rate in the face of a given number of additional simultaneous users of the same bandwidth. Each user accessing the same bandwidth in a spread-spectrum system contributes a noise level proportional to its power to the other users of the same band.

The basic method of transmission using spread spectrum depends on sending a fairly long pseudorandom sequence at a clock (or "chip") rate of 5 MHz; then this pseudorandom sequence is modulated by the data bit stream. The spreading is accomplished essentially by sending a given number of chips of the pseudorandom sequence for one bit of the modulating information sequence. The resulting signal is a pseudorandom sequence occupying 5 MHz of bandwidth which is phase inverted in consonance with the modulating bit stream. At the receiving end a copy of the pseudorandom sequence is convoluted with the incoming signal, which then reproduces the information bit stream.

The carrier-to-noise ratio of the receive signal in the 5-MHz bandwidth is generally a negative number in decibels. The processing gain, which is the number of chips per bit of the modulating information sequence, produces a positive carrier-to-noise ratio for the modulating bit stream.

In the case of Equatorial Communications, the receive-only antenna beamwidth is about 9° and the sun has a diameter of about 0.5°. The cold space dilutes the noise temperature of the sun by more than a factor of 1000. When the sun shines directly into one of the receiving dishes, it contributes only 1 dB of additional noise to the carrier-to-noise ratio, which turns out to be insignificant. Therefore, Equatorial can offer a customer 24 hours a day, 365 days per year of service without sun outages.

The spread-spectrum transmission mode also has the ability to counteract interfering signals that might arise, for example, from intermodulation products from other signals within the transponder. The signal correlator at the receiving end transforms the spread transmission into a concentrated signal with a high carrier-to-noise ratio. At the same time, it will spread an interfering signal which might have been concentrated at some frequency in the band of the spread-spectrum signal and reduces the interfering signal to a low-level, thereby reducing its effect on the desired signal.

With the advent of relatively inexpensive microprocessors, the problem of performing the spread-spectrum signal generation and reception has become minimal and costs for that equipment have become very low.

Reference (1) provides information on other multiple access methods that may see use in the future.

6.2 HOW IS THE USER CHARGED FOR SERVICE?

When a microterminal sends a message to the hub, it attaches various messages as part of the transmission protocols. In most instances, the network protocol contains header information that identifies the calling and called departments. In some instances, additional charge data (for example a shop order number) may be added to assure that the charges are assessed in a fair manner. The hub (through which all communications must flow) is a logical location to collect and assess these charges.

In some organizations telecommunications (e.g., teleconferencing) will be billed to the using organizations at less than cost to make more efficient use of transponders. The remainder of the cost is recovered from keeping executives at their desks, where they are more productive than on an airliner. Business response times are also expedited. In our discussions with users no one mentioned any difficulty in spreading the costs.

Since communications must flow via the hub, any signals with invalid headers may be denied connections. The hub can also assign priority levels in case the outbound links become congested.

6.3 NETWORK PRICE AND COSTING POLICY

The pricing policy of microterminal networks are as diverse as the networks themselves. A number of networks are in operation and their price structure is known. Vendors will provide equipment prices and AT&T has indicated its tariffs for a proposed K_u-band small terminal system, Skynet.

6.3.1 AT&T Skynet Star Network

Tariffs for the K_u-band Skynet service for "voice" networks offer one- and two-way data and one-way voice for $300 to $400 per month per earth station location, including design, management, maintenance, and transponder capacity for a 7-year lease term. The cost of the microterminals and the hub station is a separate item to be determined on a case-by-case basis.

Two-way data transmissions between a hub station and a small station would be:

Cost per Month	Data Rate
$300–$400	9.6 kb/s
2,100	56 kb/s
52,500	1.5 Mb/s
and into 1.2-meter dishes:	
$875	9.6 kb/s
5,250	56 kb/s
131,250	1.5 Mb/s

Telecom General Corporation indicated that their KuNET services would cost about $350 per month per earth station for data rates between 1200 and 9600 b/s (see also Section 4.2).

6.3.2 Tariffs and Charges

Microterminal network vendors do not have a uniform charging or tariff structure. It is possible to ascertain space segment costs from the satellite operators. At this time, space segment is available from satellite owners on the basis of a 1-, 2-, 5-, and 7-year leases with transponder capacity sold as one, three-fourths, half, and one-eighth transponders. Even smaller increments are available.

There are two broad classes of satellite transponders available. One class includes GTE and the other includes GE. The GE (formerly RCA) transponders have approximately twice as much (3 dB more) radiated power. The charges reflect this difference. The GE-type transponders range from $250,000 to $600,000 per month. The charges for the GTE-type satellite transponder range from $150,000 to $350,000 per month. These are the charges for an entire transponder. Most microterminal networks need only a fraction of a transponder for simple data verification. As the tasks get more sophisticated and the number of stations grows, multiple transponders may be needed. Multiple transponder discounts may be negotiated.

There are several ways of marketing satellite communications network services. The first is where the customer purchases the small earth stations, the hub station, and acquires some space segment capacity. For such a service, the customer would be responsible for the operation and maintenance of the network and would amortize the cost over the number of years of useful lifetime.

Another form of marketing is for an equipment vendor or network operator to offer the customer a service on a recurring charge per month basis. The customer does not own either the earth stations or the hub station but simply for a monthly telecommunications service at a specified rate. There may also be a use (per minute or per bit) charge for each terminal.

The first form of operation is suitable for those companies that are willing to shoulder the entire burden of acquiring, owning, and maintaining the communications network. They have full control over the service and its evolution.

In the second method, the customer does not own any part of the communications network but simply leases the service on a monthly basis. There may be a large discontinuation charge to discourage short-term leases and to recover the fixed costs. Such service is, therefore, more suitable to smaller companies or to customers that do not want to incur the outlay for initial investment in a satellite communications network. This approach may be less flexible to changes in the network use or size.

Costs of the microterminals can vary considerably from vendor to vendor, and clean-cut comparisons of costs are difficult. The quotes may include different equipment and may (or may not) include installation. In some cases, the equipment is provided on a turnkey basis. Additional costs may be encountered for shipping, site preparation, and wiring within the customer's premises.

The costs for microterminals range from $6000 to $8000 apiece for some to between $12,000 and $16,500 apiece for others. The cost of a hub station can be anywhere from $700,000 to $2.4 million, depending on size and complexity of hub stations.

As the hub stations are manned, the expense for operation and maintenance of the hub station must be included in the system cost. Since the hub station investment is considerable, a shared use of the hub station among a number of networks suggests itself. One example of such a shared system is the Equatorial Communications systems where the Novanet shares the California hub station and sells its own service to its own customers.

The cost per month per remote earth station depends on the number of small earth stations within a given network and how many networks share the hub station. This results from the hub station being amortized over a greater number of small earth stations. It is estimated that from 100 to 500 small earth stations are required to properly amortize a hub earth station.

REFERENCE

1. Li, V. O. K., "Performance Evaluation of Multiple-Access Networks: Introduction and Issue Overview," *IEEE Journal on Selected Areas in Communications*, Vol. SAC-5, No. 6, July 1987.

7

OVERVIEW OF THE U.S. MARKET

7.1 POTENTIAL SIZE OF THE MICROTERMINAL MARKET

7.1.1 Background Information

Business telecommunications is the principal source of microterminal users. Government uses are a small but growing market. We examined some historic market demographics statistics of business telecommunications and found that the total business telecommunications can be divided into five groups. The groups were based on user expenditures for telecommunications services exclusive of telecommunications equipment (see Table 7.1). Each group was selected on the basis of spending approximately the same total amount on telecommunications.

The number of companies in each group is shown in Figure 7.1. The first grouping (which accounted for approximately 20% of the business telecommunications charges) contained only 49 companies. The second 20% contained the next 200 companies. The third quantile contains 1295 companies. This indicates that approximately 1500 companies (or 0.05% of over 3 million companies in the United States) use 60% of the business telecommunications. The remaining 99.95% make up the rest. This illustrates that a relatively few companies pay for most of the service. These quantiles are examined further in Figures 7.2 through 7.4.

As might be expected, the average telephone bill per company in the first quantile was substantially larger than the others (approximately $50 million). In the fifth quantile there are over 2,900,000 companies paying an average of approximately $1000 per year. These are small organizations or companies in which telecommunications does not play a major role.

TABLE 7.1   Business Telecommunications[a]

Annual Telephone Expenditures	Percent of Total Business Telephones	Number of Companies	Average Bill[b]	Employees	Cost per Employee per Year	Total Charge per Group[b]
Over $20.7 M	18.7	49	$48.4 M	134,800	$359.05	$2.37 B
$6.47–20.7 M	18.7	200	11.9 M	39,000	305.13	2.38 B
$160,000–6.47 M	18.4	1,295	1.8 M	7,800	230.77	2.33 B
$3000–$160,000	21.1	89,296	30,000	98	306.12	2.68 B
Under $3000	23.1	2.94 M	1,000	4	250.00	2.94 B

[a]M, million; B, billion.
[b]Annual, 1978.

The telecommunications cost per employee fluctuates over a comparatively narrow range ($230 to $360), as shown in Figure 7.3. As might be expected, the number of employees changes drastically within the five business groups. The first group employs over 100,000 people per company, whereas the last quantile has an average of only 4 employees per establishment.

It would seem that the greatest market opportunities for microterminals is in the top 50 to 250 business organizations in the nation. In many instances this has been the case, but small, data-intensive, and highly competitive or-

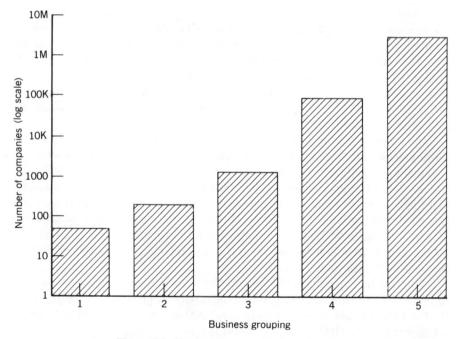

Figure 7.1. Number of companies in each group.

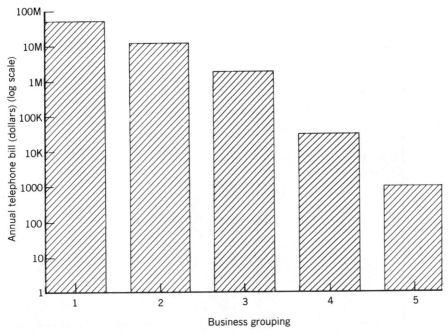

Figure 7.2. Average phone bill in each group.

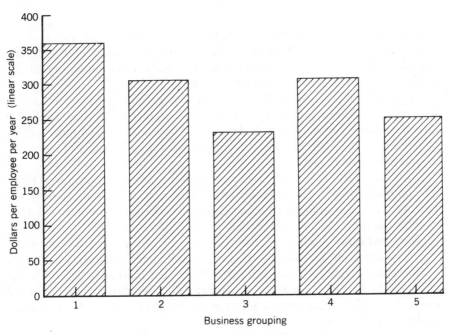

Figure 7.3. Telecommunications cost per employee.

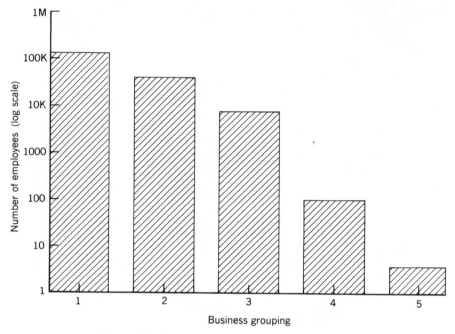

Figure 7.4. Average number of employees per group.

ganizations have used microterminals to achieve operating advantages over their larger competitors. We foresee the top several hundred retail and manufacturing organizations embracing microterminals, often on an urgent, reactive basis to counteract a competitor's move.

Eventually, the market will be the top three quantiles (about 1500 organizations). As the market moves toward these smaller users, the number of microterminals per network will decrease. The use of shared hubs will be mandatory to maintain the cost advantage. An active business of operating shared hubs will emerge.

There are various listings of commercial organizations prepared by financial magazines in the United States. The most commonly used is the Fortune 500 list of manufacturing organizations. Unfortunately, this list, by itself, is not fully adequate since there are large opportunities in the banking, service, and other nonmanufacturing industries. We predict that a similar situation will occur in Europe and other highly competitive societies.

Basically, Figures 7.1 through 7.4 indicate that marketing should be focused on the first three quantiles. Each potential candidate should be ranked on the basis of its telecommunications needs, location of its facilities, amount of data, and plans for expansion.

Local organizations (with facilities contained within a circle of 100–200 miles or 160–320 km and using low-speed data) may be poor candidates for this type of service because the economic trade-off of satellites favor longer distances. A user that is rapidly expanding (such as the Wal-Mart stores, which are constructing over 100 new stores per year) may have opportunities for microterminals even if the distances are limited. In this case, the rapid deployment of microterminals (particularly now that there is no regulatory delay for licensing) is very attractive.

7.1.2 The Growth of Digital Data

As shown in Figure 7.5, digital data represented 20% of the total telecommunications data (that was nonvideo) in 1985. Assuming that voice continues to grow at 6% per year and data at 40% per year, 1990 will be the year in which the two services have equal volumes. Beyond that point, data begin to overshadow the voice service. By 1990 the total volume will be twice 1985's level.

We believe that the 40% per year growth rate for data may be conservative and the point of equal traffic nationally in the United States may actually occur in 1989. In at least one city, this point was reached in 1986. As can be seen from this figure, the mixture of traffic is changing rapidly. In many respects, this is too rapid for the local telecommunications network (which is voice-based) to handle comfortably, and this is another reason why microterminals have an excellent opportunity to grow. Since microterminals

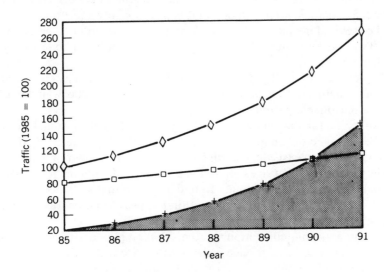

□ , Voice; +, data; ◊ , total

Figure 7.5. Data volume growth.

handle digital data, they are a natural method of handling the increasing digital traffic volume.

The local exchange carriers are making new local loop installations in fiber (instead of copper) wherever possible. Since 80% of the copper in telephone service is in the local loop, it may be many years before many business locations receive the benefit of fiber. Southern Bell has the ambition to replace all copper by the year 2000.

7.2 GROWTH ENCOURAGEMENT FACTORS

There are a number of factors that encouraged the growth of the microterminal network market. One of these was the FCC ruling on licensing of small K_u-band two-way earth stations. It allowed vendors or users to obtain one license for each *type* of small earth station (instead of one per station) and a separate license for the network hub station. The license application is a two-page form with one page for technical information and one for financial data.

Whereas the simple function of the licensing process is not in itself a major factor in encouraging growth in small terminal satellite networks, it does interact with another factor, timeliness of installation.

The Equatorial SSMA C-band transmit stations still require licensing, but may be installed on a noninterference basis and are type accepted. If they create interference with a terrestrial microwave link, they must shut down.

One of the selling points of small terminal satellite networks is that a customer can order a service and obtain it in a reasonably short time—we have heard of several instances of 2 days from placing the order to obtaining service. Two weeks is more likely and is many times faster than the typical telephone company.

The relaxation of the licensing for K_u-band terminals enhanced the timeliness factor and encouraged potential customers who wanted service within a reasonable period.

Another factor that encouraged the growth in small terminal networks was the entrance of a number of vendors into the K_u-band field. It is now possible to buy or lease small earth terminals from a number of vendors with a variety of network services offered. This choice of equipment and services will make it easier for a customer to find a system that suits his needs, even though the variety of offerings may at first appear to be confusing and perplexing.

Satellite data communications systems that employ earth terminals at the customer's premises can provide high-speed service with very low error rates, which is difficult to obtain via normal telephone lines for speeds higher than 9.6 kbps. These are ideal for customers requiring higher-speed data service, virtually error free, and looking toward satellite communications, particularly if they also need to service many diverse locations.

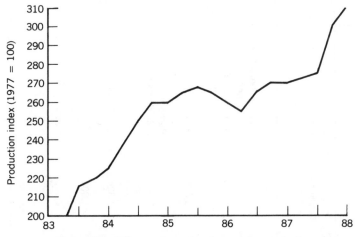

Figure 7.6. Office equipment and computer production.

In the mid-1980s the terrestrial network was still basically an analog network with pieces of digital service. As such, ISDN may be closer in space than it is on the ground. For data users the microterminal will provide higher-quality service, may be installed quickly, and may be moved at will while providing the data services necessary to collect or process information.

One of the most attractive aspects of the microterminal is that it presents stable charges in an unstable world of telecommunications costs. Once the earth station is installed, the only potentially variable cost is for the space segment. This portion may also be stabilized by the purchase of one or more transponders or entering into a long-term lease agreement.

Figure 7.6 shows the 12% annual growth of office equipment and computer production in recent years. Both the past and future history suggest that business is becoming increasingly digital, and therefore has the equipment necessary to make a microterminal network a success. Most of the existing digital office equipment (such as word processors and personal computers) tend to operate either on a stand-alone basis or within a very limited office network (called a local area network or LAN). Some LANs are beginning to form connections with other LANs, both within the same corporate organization and to outside suppliers and customers. The microterminal easily permits further opportunities in this direction, especially if the hub is shared by more than one user.

7.3 ECONOMICS AND MARKET SIZE

The market penetration for these terminals is based on two separate markets. In previous years it was believed that microterminals would only be attractive

TABLE 7.2 Microterminals in Segments of U.S. Labor Force

Occupation	US Civilian Work Force			People per Terminal (×1)	Terminals in 1992 (×1)	Uses
	1970 (×1000)	1980 (×1000)	1980 (%)			
Executive and administrative	5,970	10,379	10.0	300	34,597	Remote computations and searches
Professional	8,800	12,275	11.8	50	245,500	
Technicians and related support	1,821	3,063	3.0	200	15,315	Remote data
Salespeople	8,021	10,257	9.9	250	41,028	Department stores
Administrators	13,207	17,564	16.9	200	87,820	
Service occupations	10,186	13,606	13.1	1,000	13,606	Repair data, parts and gas stations
Farming, fishing and forests	3,034	3,032	2.9	50,000	61	Rural communications
Mechanics and repairers	3,275	3,983	3.8	500	7,966	Auto dealers
Construction trades	3,615	4,814	4.6	3,000	1,605	
Extractive	177	314	0.3	200	1,570	Oil
Precision production	4,091	4,326	4.2	1,000	4,326	CAD/CAM
Woodworking	110	118	0.1	90,000	1	
Machine operators and assemblers	8,938	10,082	9.7	10,000	1,008	
Transportation and moving sat'l	3,932	4,820	4.6	2,000	2,410	Inventory
Handlers, cleaners, and laborers	4,515	5,086	4.9	90,000	57	
Totals	79,692	103,719			456,869	

to large corporations, and therefore, the number of people per building was used as a determinate on market penetration.

Although this technique is still valid, it is now obvious that there is at least one other major market for K_u-band stations. This is at the retail level. Both department stores (employing hundreds of people) and gasoline stations (employing perhaps 10 people) are now included in the potential market for retail earth stations.

Table 7.2 lists the major segments of the U.S. labor force in 1970 and 1980, the percent of the work force in each category, and an estimate of how many people would be needed to support a microterminal.

Different skill levels and occupations have more or less need for a digital data terminal. Woodworkers often work alone or in isolated groups with little need for outside contact (hence one terminal per 90,000 workers). Professionals (engineers, computer scientists, teachers, lawyers, doctors, etc.) have a far greater need for the power of remote computation, data base accession, and so on. They also tend to be in small groups, hence the 50 professionals per terminal. Considering that the terminal cost is less than an automobile, the 50:1 ratio may be conservative. Professionals in industry could easily use $\frac{1}{50}$ of a microterminal at the busy hours.

Several occupations may contain surprises. Sales include small convenience stores such as the 7-Eleven (12–20 employees) to large department stores like K-mart (100–300 employees). We have estimated one terminal per 250 employees because there are so many small retailers.

Service occupations include gasoline stations (credit verification) and repair depots (which may need facsimile access to the repair manuals at the manufacturer's plant and the ability to order parts). The mechanics category includes, among others, automobile dealers (who are on the way to an inventory control system).

These numbers suggest a half million terminals in 1992.

Frost & Sullivan predicts the $1.2 billion spent in 1983 for business data communications will reach $2.3 billion in 1986 with a growth rate of 38%. If this continues at this pace, it will reach $42 billion per year by 1995. Between 1986 and 1995, $143 billion will have been spent in this area. Other studies by the Communications Center suggest that the microterminal and hub expenditures may reach $4.3 billion in this time period. This represents only 3% of the F&S estimate for all business data communications.

7.4 T1 INSTALLATIONS

In addition to the 56/64-kb/s microterminal, there is a market for T1/E1 terminals running at 1.544 and/or 2.048 Mb/s. E1 is the T1 equivalent in most of the rest of the world outside North America.

These become attractive in the United States for terrestrial service when 10 of their 24 voice circuits can be filled due to the quantity advantage of capacity in bulk instead of by the individual circuit. Reconfigurable arrangements and some limited switching is now available for user control. The rate of the T1 growth in the United States is 20–40% per year (depending on different parts of the country) in the 1980s.

Terrestrial T1 circuits require several months to install and are expensive to operate. If they can carry voice they are subject to access charges equivalent to 24 voice circuits.

T1 earth stations are available for one-time charges of $15,000–$25,000 each. The quantity sold to 1987 was small.

We estimate that as the popularity and acceptability of the 56/64-kb/s stations grow, there will be a rapid deployment of T1 stations after certain technologies become available.

They are:

1. Higher-powered solid-state 14-GHz power amplifiers (20–50 W)
2. Equipment for the hub to demultiplex the T1, switch the individual data streams to the proper destination beams, and remultiplex outgoing T1s

TABLE 7.3 Per Site Installation Costs, 1987

Data rate:	56 kb/s	T1
On-site time[b]	3 hours	8 hours
Sites per day	2	1
People per field crew	2	2
Rate/day[a]	$ 200	$ 200
Rate/site	$ 100	$ 200
Truck and equipment		
($35,000, 5 years)	$ 18	$ 35
Expenditures	100	100
Shop support (one person)	50	100
Spare parts for installation	50	100
Gas, lodging, etc.	100	200
Subtotal direct	418	735
Overhead (benefits, vacations,		
etc.) 120% of labor	180	360
Subtotal	598	1,095
General and administrative (10%)	60	110
Subtotal indirect	240	470
Total[c]	838	1,205
Basic equipment costs	$10,000	$25,000
Percent of basic costs	8.4	4.8

[a]Two people at $20,000 per year.
[b]Microterminal sites with reasonable access.
[c]Before profit

TABLE 7.4   Maintenance Contracts, 1986–1987

Item	Initial Cost	Maintenance		Annual (% of Initial)
		Monthly	Annual	
Word processor	$30,000	$197	$2,364	8
Laser printer	5,000	74	888	18
Copier	8,000	—	595	7
Sears' appliances	—	—	—	10

3. Sharing of the hub by many users to make the T1 service available and affordable

4. Computers and other equipment designed to use T1 rates

The T1 market is presently centered on the larger users identified in Section 7.1.1 for voice-multiplexed data and teleconferencing. Eventually, bulk digital applications (such as CAD/CAM) will evolve. We estimate that for every 10 K_u-band 56-kb/s terminal, there will be one T1 terminal in 1995. The space segment requirements are limited in the same manner as the 56-kb/s service except that about 30 times (or 15 dB) as much power and bandwidth are needed per carrier.

7.5   INSTALLATION OF MICROTERMINALS

Table 7.3 shows the components of the installation costs. Some installation crews use former cable television and backyard satellite (TVRO) crews that may make less money (10–15%) and have fewer benefits. Callbacks may consume the apparent savings if the installation is not done properly.

7.6   MAINTENANCE OF MICROTERMINALS

We examined a number of consumer and business equipment service agreements for equipment maintenance. These fell into the annual range (8–18%) of the initial price of the item, with 10% being the most common (see Table 7.4). We feel that a microterminal is less trouble-call prone than a laser printer (which has moving parts and high technology), more troublesome than a word processor, and at least as well protected as consumer products. We have selected 10% per year (0.8% per month) as a typical maintenance cost.

NETWORK OPERATORS

8.1 COMMON CARRIERS

Up until the early 1970s the description of common carrier was very simple. It was a communications company that offered its services to all customers on a nondiscriminatory basis at a uniform price schedule that was regulated by the government. At the national level there were two major carriers: American Telephone and Telegraph (AT&T) for telephone service and the Western Union Telegraph Company (WU) for telegrams. Within individual states there was between one and several tens of individual companies, all providing services within carefully prescribed franchise areas. The largest number of these statewide carriers were part of the AT&T family ("The Bell System"). The General Telephone System (GTE) had the second largest network of local exchange telephone companies.

Competition emerged first in the international record carriers (originally telegrams and later video and voice services). This competition grew out of individual commercial companies that had been established to provide radio-telephone services to ships and handle other overseas telecommunications. It spread to the establishment of domestic long-distance carriers.

8.2 SATELLITE CARRIERS

The advent of the satellite for telecommunications was accompanied by a long debate within the U.S. government as to whether the satellite services should be provided exclusively by AT&T or whether competition should be permitted. This result was the so-called "open skies policy," which allowed any qualified organization to offer satellite services.

The Commission was encouraged by the wide variety of new services offered by the competing satellite operators (which in the beginning were RCA, Western Union, American Satellite, and SBS). These are now GE, WU, Contel-ASC, and MCI, respectively, due to various acquisitions. Table 8.1 lists the principal satellite common carriers in the United States and their satellites.

Most of the organizations are self-explanatory, but three may require further explanation for a full understanding. Alascom is the carrier for Alaska. Satellite Business Systems was originally established as a joint venture of Comsat (the Communications Satellite Corporation), IBM (the computer manufacturer), and Aetna Life and Casualty (an insurance company which had assets nearly equal to those of IBM). Due to a series of early decisions, the satellite system required large (5.5 and 7.7-m) earth stations at K_u-band. IBM provided each earth station with the controller. This combination proved to be expensive and the SBS system did not break even under the joint venture. Eventually, first Comsat and then later Aetna Life and Casualty sold out to IBM. In 1985, IBM transferred most of its ownership in SBS to MCI in return for an ownership share of MCI. IBM retained ownership of three satellites (one launched and two to be launched). In 1986 IBM Transponder Leasing Corporation was established to handle SBS-4.

The early managers of SBS opted for a simple single-beam approach which required the large stations. Another conservative decision was to place only 10 transponders on the satellite and use a single polarization. The Canadian Anik-C series (procured simultaneously with SBS) used both polarizations and a full 16 transponders. SBS, therefore, carried the double penalty of excessively large and costly earth stations and only half the space segment capacity per satellite, which resulted in a higher per transponder cost.

The second principal satellite common carrier was the American Satellite Company. This was originally founded by Fairchild Industries. Later, it be-

TABLE 8.1 Principal Domestic U.S. Satellite Common Carriers

Carrier	Parent
Alascom, Inc.	Pacific Telecom
Contel-ASC	Contel
AT&T Communications	AT&T
Equatorial Communications Company	Contel-ASC
GTE Spacenet Corporation	GTE
Hughes Communications Inc.	GM
IBM Transponder Leasing Corporation	IBM
MCI Communications Inc. (SBS)	—
GE American Communications (RCA)	GE
Western Union Telegraph Company	—

came a joint (50%–50%) venture with Continental Telephone (Contel). In 1985, Continental Telephone acquired 100% of the American Satellite Company and eventually renamed it Contel-ASC. Contel-ASC owns 20% of the Westar System (*Westar III–V*), has launched one satellite of its own (*ASC-1*), and has a second satellite under construction for future launch. Each of the major terrestrial carriers (AT&T, General Telephone, MCI, and Continental Telephone) are represented among the satellite common carriers.

In the late 1980s another form of carrier emerged, the telehub (sometimes associated with a teleport) operator. These organizations provided shared hub services to multiple microterminal networks. In some cases they emerged from a corporate network that had constructed its own network and saw a business opportunity in offering its services to others. Profits were seen in the difference between the incremental cost of adding capacity to an existing hub and the cost of an entire telehut. Other operators established telehubs in the hope of attracting new customers (which is the same practice employed in new fiber optics routes, teleports, and other telecommunications ventures).

9

THE ECONOMICS OF MICROTERMINAL VERSUS TERRESTRIAL SERVICES

9.1 THE TERRESTRIAL NETWORK

At one time, there was one major telephone company in the United States (American Telephone and Telegraph). AT&T was responsible for the entire network, including the local loops, the local telephone exchanges, interexchange links, toll switches, the long-distance terrestrial network, the Comstar/Telstar satellites, and international traffic.

To accommodate all of these functions, AT&T devised a hierarchy for switching traffic. The telecommunications highways included coaxial cable, microwave, and some satellite links. The network included all the switches, protocol equipment, and network maintenance. It set the standards for its customers.

The operating divisions are shown in Figure 9.1.

9.1.1 The Post-Divestiture Situation

In 1984, the operating part of the Bell System was broken into AT&T Communications and the seven Regional Bell Operating Companies. The RBOCs, in turn, controlled other companies, including the local exchange carriers (LECs; see Figure 9.2). The territories of each of the Regional Bell Operating Companies was broken down further into what became known as local area transport arrangements (LATAs). Within a LATA the Regional Bell Operating Company's LEC subsidiary was allowed to provide local exchange service. If the communications crossed a LATA, then AT&T Communications provided the link. During this period the other common carriers (MCI,

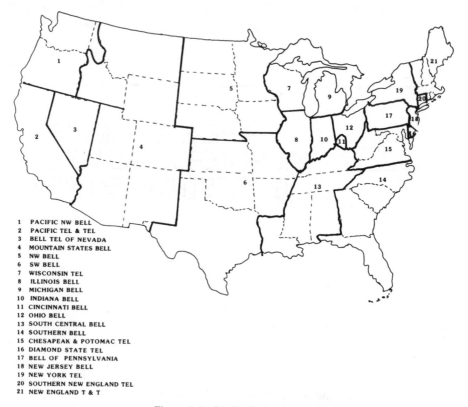

1   PACIFIC NW BELL
2   PACIFIC TEL & TEL
3   BELL TEL OF NEVADA
4   MOUNTAIN STATES BELL
5   NW BELL
6   SW BELL
7   WISCONSIN TEL
8   ILLINOIS BELL
9   MICHIGAN BELL
10 INDIANA BELL
11 CINCINNATI BELL
12 OHIO BELL
13 SOUTH CENTRAL BELL
14 SOUTHERN BELL
15 CHESAPEAK & POTOMAC TEL
16 DIAMOND STATE TEL
17 BELL OF PENNSYLVANIA
18 NEW JERSEY BELL
19 NEW YORK TEL
20 SOUTHERN NEW ENGLAND TEL
21 NEW ENGLAND T & T

Figure 9.1. Original Bell System.

Sprint, etc.) gained momentum with their own nationwide and trans-LATA networks using a combination of terrestrial microwave, leased satellites, and leased AT&T circuits. Later they either constructed or leased fiber optics. In addition to regular telephone services, another series of common carriers provided value-added networks (primarily using packet switching).

As we have seen in Chapter 7, a small minority of large companies and service organizations use and pay for the majority of the services. These organizations quickly understood that they could negotiate with the value-added networks and other common carriers on a contract basis rather than being dependent on tariffs, which were constantly being escalated upward. Many companies constructed their own nodes and even private microwave and fiber optics networks. These networks evolved into a combination of private and public facilities, thereby diverting some revenue from AT&T and the RBOCs. This resulted in higher prices for everyone else, which further accelerated the trend.

The divestiture left AT&T with the most productive parts of the network (long-distance and trunk interexchange facilities) and the RBOCs with the

Figure 9.2. Post divestiture.

labor-intensive part (stringing copper wires on wooden poles and operating the local exchanges). The local exchange companies had been dependent on the long-distance service to subsidize their operations. With time the subsidy diminished even with the revenues that the local exchange companies obtained from access circuits in return for supplying the local connections to the long-distance networks (either AT&T or other common carriers). These access charges provided further incentives to construct private networks for large organizations.

At the same time, the end user was permitted to select between the telephone company and other manufacturers for telephones, modems, and so on. While the equipment had to be certified that it met FCC requirements before it could be connected to the telephone system, this was not an overwhelming burden. The net result was a massive change in the distribution of ownership of telephone equipment, lines, and facilities. Not only was it moved from AT&T to the RBOCs but also through private industry as it began to take over supplying its own telecommunications.

This shift enabled corporate America to take control over some of its own telecommunications destiny. Competition in America has changed drastically over the last 15 years. More and more, the United States became dependent on information and its flow. Anything that slows down or obstructs the flow affects the competitive nature of American business. This information flow was originally thought of as being primarily within the various lower levels of an organization and its passage onto the top in the form of paper mem-

TABLE 9.1   56-kb/s Service Offerings, Mid-1986

Company	Name of Offering	Component	N.Y.–L.A.[a] (2443 miles)	Wash.–N.J.[a] (195 miles)
AT&T	Dataphone digital service (DDS)	Interexchange miles	$ 28,583/mo	$ 4,563/mo
		Service connections (2)	1,874/mo	1,874/mo
		Installation[b]	650	650
		Station terminals	39/mo	39/mo
		Installation[b]	156	156
		Per month	$ 30,496	$ 6,476
		Installation	806	806
		5-year total[c]	$1,830,566	$389,366
	Accunet packet service	Internodal trunks	$ 28,583/mo	$ 4,563/mo
		Node facilities (2)	2,000/mo	2,000/mo
		Installation[b]	2,000	2,000
		1000 kilopackets/mo at 0.82	820/mo	820/mo
		Per month	$ 31,403	$ 7,383
		Installation	2,000	2,000
		5-year total[c]	$1,886,180	$444,980
	Private lines	Interoffice channels	$ 52,338/mo	$ 1,256/mo
		DDS central office: connection—type 2(2)	476/mo	476/mo
		Installation[b]	896	896
		Local portion (type 2): interrate centers	(assumes 5 miles)	
		New York	424/mo	—
		California	624/mo	—
		Washington, D.C.	—	—
		New Jersey	—	—
		Access coordination	54/mo	54/mo
		Installation[b]	260	260
		Per month	$ 53,916	$ 2,664
		Installation	1,156	1,156
		5-year total[c]	$3,236,116	$160,996
	Skynet C-band service (56 kb/s)	Space segment	$ 1,860/mo	$ 1,860/mo
		Earth segment		
		Fixed antenna (2)	6,260/mo	6,260/mo
		Transmit terminals (2)	1,240/mo	1,240/mo
		Installations (2)[b]	12,086	12,086
		Receive terminals (2) .	725/mo	725/mo
		Installations (2)[b]	12,086	12,086
		Per month	$ 10,094	$ 10,094
		Installation	24,172	24,172
		5-year total[c]	$ 629,812	$629,812
	Skynet K$_u$-band service (56 kb/s) (full-time service)	Transmit/receive stations	(not included)	
		Hub station	(not included)	
		Hub services	(not included)	

TABLE 9.1 Continued

Company	Name of Offering	Component	N.Y.–L.A.[a] (2443 miles)	Wash.–N.J.[a] (195 miles)
		Space segment		
		Hub to 1.8-m station		
		(1) at	$ 900/mo	$ 900/mo
		[Hub to 1.2-m station]	[$ 10,500/mo	$ 10,500/mo]
		From 1.2- or 1.8-m		
		station to hub (2) at	4,200/mo	4,200/mo
		Local channel services	(not included)	
		5-year full-time space		
		segment	504,000	504,000
		2 microterminals		
		(installed)	30,000	30,000
		Hub use (estimated)	20,000	20,000
		5-year total (estimate)	$ 554,000	$554,000
Argo	Argopoint 56-kb/s digital service (for reference only, see text)	Earth station satellite	$ 4,200/mo	$ 4,200/mo
		Installation[b]	100	100
		Local access line:		
		interrate centers	(assumes 5 miles)	
		New York	424/mo	—
		California	624/mo	—
		Washington, D.C.	—	244/mo
		New Jersey	—	634/mo
		Access coordination	54/mo	54/mo
		Installation[b]	260	260
		DDS central office:		
		connections—type		
		2 (2)	476/mo	476/mo
		Installation[b]	896	896
		Per month	$ 5,788	$ 5,608
		Installation	1,256	1,256
		5-year total[c]	$ 347,936	$337,736
MCI	Private line (56 kb/s)	Interoffice channel	$ 4,647/mo	$ 1,058/mo
		Central office		
		connections (2)	80/mo	80/mo
		Installation (2)[b]	402	402
		DDS multiplexer (2)	100/mo	100/mo
		Local access (2)	54/mo	54/mo
		Installation (2)[b]	468	468
		Local access surcharge		
		New York	36/mo	—
		California	35/mo	—
		Washington, D.C.	—	41/mo
		New Jersey	—	31/mo
		Per month	$ 4,953	$ 2,235
		Installation	870	870
		5-year total[c]	$ 298,050	$134,970

[a]Typical routes with distances per the FCC tariffs.
[b]One-time charge.
[c]Sum of the one-time and monthly charges.

orandums and the like. Now much of the information is keyed in to various data bases which are now accessible at all the levels of the organization. The president of a company, for example, can determine how many sales were made last week, what types of customers complaints are being received, and how long it takes to rectify each complaint. Now, having information flow within a building is not enough. It is desirable to be able to access any information in any corner of the country. This requires extensive telecommunications to link all of the various elements together on either a star or a switched mesh basis. Equipment is now available from private sources to do the functions that formerly only AT&T was allowed to do and had the necessary equipment to do.

Some organizations with far-flung operations (such as coal mining, liquid petroleum pipelines, oil rigs, etc.) felt that their operations were being impaired by the remote locations and the limited telecommunications available from the traditional telephony companies. Out of this has come some private microwave and a potential market for microterminals.

Unlike the predivestiture era, AT&T no longer has control over all parts of the network. The most troublesome part (the local loops) is no longer part of their network. They, like the corporate users, must now deal with at least seven regional holding companies (and in some cases, independent telephone companies which are scattered throughout the United States).

Out of this background of the terrestrial network comes the opportunity for the small earth terminal.

9.1.2 Typical Cost Components

Table 9.1 shows the cost elements of five AT&T Communications tariffs and three other common carrier tariffs. As can be seen from Figure 9.3, there is a wide variation in the cost of providing 56-kb/s services, depending on the carrier and the particular tariff involved. Two circuits (New York to Los Angeles and Washington, DC, to northern New Jersey) were selected as examples of long-distance and medium-distance traffic. As shown in Table 9.2, the New York-to-Los Angeles service is 2443 miles between rate centers. In the case of the Washington-to-northern New Jersey path, it is 195 statute miles.

As can be observed from Figure 9.3, the Argo and AT&T satellite services offer relatively low costs for the New York-to-Los Angeles link. Argo has gone out of business. All tend to have a distance-insensitive price, as should be expected from satellites. The two services compete favorably with terrestrial AT&T offerings. MCI's prices are the lowest but still total nearly $300,000 for the long-distance case and $135,000 for the short-haul example.

Two AT&T satellite services (at C- and K_u-based) are provided for comparison with the terrestrial counterparts.

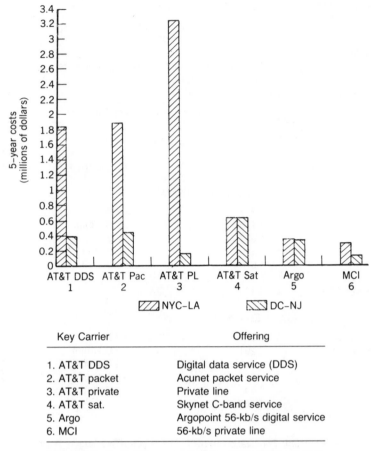

Figure 9.3. Cost of 5 years of service (FCC 56-kb/s tariffs).

Key Carrier	Offering
1. AT&T DDS	Digital data service (DDS)
2. AT&T packet	Acunet packet service
3. AT&T private	Private line
4. AT&T sat.	Skynet C-band service
5. Argo	Argopoint 56-kb/s digital service
6. MCI	56-kb/s private line

9.2 THE SPACE SEGMENT

9.2.1 Leased Space Segment

The total 5-year estimate for the 56-kb/s full circuit using the AT&T Skynet K_u-based services is $554,000 regardless of distance. This makes the AT&T microterminal cost competitive on the New York-to-Los Angeles path for all of the other AT&T offerings. It is still above the MCI price. For the short-distance (Washington to New Jersey) path, this price is generally above the nonsatellite offerings.

The installed cost of just two microterminals is $12,000–$30,000 (we will take the $30,000 value). Another $20,000 is allotted for the 5-year use of the

TABLE 9.2   AT&T 56-kb/s Satellite Service (C-band)

Provided via:	Telstar 300 series of C-band satellites
Basis:	Single channel per carrier (SCPC)
Antenna diameter:	6.1 meters (smallest size offered)
Price basis:	AT&T FCC Tariff 7, Section 7, Skynet Digital Service
Revision used:	July 3, 1987 (Effective date)
Term of service:	3 years

	Cost per month	Tariff Paragraph	Tariff Page
	Space Segment		
	$ 1,800 (2 at $900)	7.5.1.A.2(a)	91.6
	Earth Segment: First Station[a]		
Fixed antenna	$ 3,374	7.5.2.B.1(a)(i)	91.8
Transmit terminal	1,135	7.5.2.B.1(b)I(i)	91.9
Installation[b]	11,670	7.5.2.B.1(b)(ii)(1)	91.10
Receive terminal	499	7.5.2.B.1(d)I	91.13
Installation[b]	11,670	7.5.2.B.1(d)I	91.13
Monthly subtotal	$ 5,008		
Installation total[b]	$ 23,340		
	Earth Segment: Second Station[a]		
Fixed antenna	$ 3,374	7.5.2.B.1(a)(i)	91.8
Transmit terminal	201	7.5.2.B.1(b)I(i)	91.9
Installation[b]	743	7.5.2.B.1(b)(ii)(1)	91.10
Receive terminal	287	7.5.2.B.1(d)I	91.13
Installation[b]	743	7.5.2.B.1(d)I	91.13
Monthly subtotal	$ 3,862		
Installation total[b]	$ 1,486		
	Earth Segment: Two-Station Network Totals (with Two Space Segments)		
Monthly total	$ 10,670		
Installation total[b]	$ 24,826		
Three-year total[c]	$408,946		
	Earth Segment: Ten-Station Network Totals (with Four Space Segments)		
Monthly total	$ 20,194		
Installation total[b]	$ 27,798		
Three-year total[c]	$754,782		

[a]On or before January 23, 1987. Individual prices will be determined thereafter (see 7.5.2.B).
[b]One-time charge.
[c]Includes 36 monthly payments and the installation charges.

hub (a one full-circuit share of the total hub's capacity). The leased space segment may be $400–900 per month per direction. The 5-year total of this approach is $158,000.

If this analysis were carried out on a mesh basis where there are many microterminals, a different result would be expected. Instead of dedicated lines between each location, a time-shared space segment would be used.

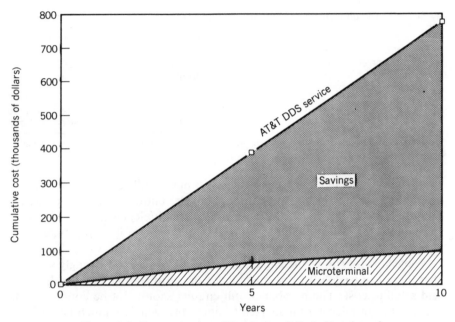

Figure 9.4. Cost comparison (Washington, D.C., to New Jersey).

The Skynet space segment ($504,000 per 5 years) may be time-shared among 17 to 25 earth stations (depending on the type and volume of traffic), thereby lowering the per earth station cost of the space segment to $2,000–$30,000 (per 5 years).

There is another difference between the purchased microterminal system and all the others shown in Figure 9.4. At the end of the 5 years, the microterminal has been amortized and is the property of the user. In all the other cases, the facilities are still the property of the common carrier. At the entry into year 6, microterminals have been paid for. Depreciation can be recovered from federal taxes. In the case of a service lease, there is no depreciation because the property does not change hands during the lease. The lease payments are deductible as a normal operating expense.

9.2.2 Purchased Space Segment

Instead of paying AT&T $504,000 for 5 years for a single 56-kb/s full circuit, it is possible to purchase a transponder outright that is capable of handling 100 full circuits (using 1.2-m stations) or 200 (with 1.8-m stations) full circuits. Typical anticipated lifetimes of these transponders are 10 years. Table 9.3 shows the circuit cost (on a per month basis) assuming that the transponder is purchased when the satellite goes into operation and is kept in full service throughout the entire 10-year lifetime. An interest rate of 10% is assumed to pay for the transponder.

TABLE 9.3 Circuit Costs if Transponder is Purchased[a]

Microterminal Antenna Diameter (m)	56-kb/s Full Circuits per Transponder	Purchase Price (millions of dollars)				
		10	14	18	20	24
		Monthly "Cost" per Full Circuit				
1.2	100	$1322	$1850	$2379	$2643	$3172
1.8	200	661	925	1189	1322	1586
2.5	400	330	462	595	661	793

[a]Assumes an interest rate of 10% per year (currently, prime rate + 1%). The transponder is at K_u-band.

In reality, the satellite transponder will not be filled on the first day of operation and remain that way throughout the entire 10 years. The actual monthly cost will be higher because of these inefficiencies. If it is assumed that, on average, the satellite is utilized to only 50% of its capacity, each of these numbers should be doubled.

Table 9.3 also shows the impact of earth station size and strongly suggests that there is a trade-off to be made between earth station antenna diameter and satellite cost. The number of full circuits should not be confused with the number of microterminals (see Chapter 14), which is much larger.

Table 9.4 examines the total K_u-band satellite network costs at a station pair level. Both leasing (from AT&T) and owning (an $18 million transponder) are considered. The depreciation of the space and ground segments is not included. The ground segment (initial cost $30,000) is financed over a 5-year period at 10% per year.

Table 9.5 compares the various approaches. The purchased microterminal emerges as, by far, the lowest-cost approach and is substantially lower than even the shorter link (Washington, D.C., to northern New Jersey). The microterminals can break even in 1–3 years compared to AT&T terrestrial tariffs. This cost saving, coupled with convenience and other factors should result in an active market for these stations. At 1.544 Mb/s the economics in favor of microterminals are even stronger because the equipment cost rises slower than the derived capacity.

9.2.3 Trunk C and K_u-Band Satellites

Table 9.6 shows the typical prices for transponders leased by the month. These are approximate values which include all of the normal space segment charges. The earth stations are to be supplied by the user and therefore are not included in the table. The C-band transponders (with the exception of spread-spectrum multiple-access stations) generally require antennas too large to be a microterminal.

If the user has many circuits, it may be advantageous to consolidate these to take advantage of some of the wider bit rate or bandwidth service offerings of the fiber and satellite carriers. When large earth stations are used along

TABLE 9.4   Satellite Network Costs

Space Segment (56- or 64-kb/s Full Circuit)	Cost	Notes
AT&T Skynet K$_u$ Lease	$504,000/5 yr	Table 9.1
(1.8-m antenna)	8,400/month	Table 9.1
Purchased at $18 million	1,189/month	Table 9.1
If used at 50%	2,378/month	

	AT&T Lease	Purchased
Time shared by:		
17 users	$494/mo/station	$140/mo/station
250 users	34/mo/station	10/mo/station

Earth Segment	Cost for First 5 Years	Cost After 5 Years	Notes
Station cost (each)	$15,000	0	Installed
Monthly payment	287/mo	0	Principle + 10% Interest, 5 yr
Maintenance and supplies	125/mo	$125/mo	10% of $15,000 (per year)
Total (per station)	$ 412/mo	$125/mo	

	AT&T Skynet K$_u$	Purchased

Total Costs per Month for First 5 Years

Space segment				
Per month	$8,400		$2,378 at 50%	
Users	17	250	17	250
Per station month	$494	$34	$140	$10
Earth segment: per 1.8-m station month	$412	$412	$412	$412
Subtotal				
Times 2 (earth stations)	2	2	2	2
Total per month	$1,812	$892	$1,104	$844
Months in 5 years	60	60	60	60
Total first 5 years	$108,720	$53,520	$66,240	$50,640

Total Costs per Month After 5 Years

Space segment: Per station month	$494	$34	$140	$10
Earth segment	$125	$125	$125	$125
Subtotal	$619	$159	$265	$135
Times 2 (earth stations)	2	2	2	2
Total per month	$1,238	$318	$530	$270
Months in 5 years	60	60	60	60
Total second 5 years	$74,280	$19,080	$31,800	$16,200

TABLE 9.5 Charges for 56 kb/s over Five Years

Carrier	Offering	N.Y.–L.A.	Wash.–N.J.
AT&T	Dataphone digital service (DDS)	$1,830,566	$389,366
	Accunet packet service	1,886,180	444,980
	Private line	3,236,116	160,996
	Skynet C-band service	629,812	629,812
Argo	Argopoint 56-kb/s digital service	347,936	337,736
MCI	56-kb/s private line	294,050	134,970
		Distance Independent	
—	Purchased microterminal with AT&T Skynet K$_u$		
	First 5 years	$53,520–$108,720	
	Next 5 years	19,080– 74,280	
—	Purchased microterminal with purchased transponder		
	First 5 years	$50,640–$66,240	
	Next 5 years	16,200– 31,800	

with TDMA or companded single-sideband amplitude modulation (CSSB/AM), the channel capacity of the transponder may be pushed into the 1000–6000 half-circuit range. This is the traditional mode of operation of these transponders at C-band and to some extent at K$_u$-band.

9.3 FIBER OPTICS

Fiber optics has a distinct advantage of being a high-capacity low per circuit cost-transmission media. At the same time they have the disadvantages of requiring access to the local switched signal network, which has been the Achilles heel of the traditional copper or microwave or trunk satellite systems. These local loops tend to be noisy and costly. The Communications Center has conducted studies for other clients of the satellite versus fiber issue. In many instances, the local distribution cost predominated the cost of using fiber optics.

In most instances, the present fiber networks have nodes located only in

TABLE 9.6 Full Transponder Leases, Mid-1987

Band	Bandwidth (MHz)	Typical Price per Month		
		Protected	Unprotected	Preemptible
C	36	$150,000	$100,000	$ 66,607
K$_u$	54 or 72	320,000	200,000	120,000
	43	180,000	100,000	70,000

the major cities. Industries located in suburban areas (and rural areas) must lease the local copper facilities to reach the fiber nodes or construct bypass (typically microwave) facilities to reach these nodes. Fiber optics does a magnificent job of high-volume traffic (the very opposite end of the market specter from the microterminals). Microterminals is a market unto itself, separate and distinct from these terrestrial services.

10

MICROTERMINAL NETWORK OPERATIONS

10.1 MICROTERMINAL NETWORKS

All microterminal networks using hubs have similar operating requirements. We have chosen one example to illustrate how these networks operate and the functions of each element.

10.1.1 Corporate Networks

The following description has been fabricated. It is intended only as one example of a multidestination data broadcast network linked by satellite. Other digital record and facsimile networks operate in a similar manner. In these cases, identical sets of information are sent to multiple addresses (branch offices, factories, etc.).

Assume that a 54-MHz transponder (from microterminal to hub) can accommodate 600 56-kb/s data channels. If the forward error correction is at rate $\frac{2}{3}$, the actual information rate will be $37\frac{1}{3}$ kb/s per second ($56 \times \frac{2}{3} = 37.33$). This coding is done at the transmitting microterminal, as it is more of a telecommunication than a computer function.

The hub can detect a weaker signal from the satellite than a remote station can, due to the higher gain in the hub's receiving antenna. Therefore, the hub-to-microterminal link must have more signal-transmitting power than the microterminal-to-hub link. For this reason, time-division multiplex (TDM) is often chosen for the hub-to-microterminal and SCPC for microterminal-to-hub communications. Figures 10.1 through 10.10 follow the hypothetical transmission sequence.

Channel Request. In Figure 10.1 the originating small station (1) sends an ALOHA request to the hub and the network operations center (NOC). ALOHA is a random multiple-access transmission method based on contention. In Figure 10.2 the NOC (9) takes notice of the request, consults a list of available inbound frequencies (8) in the transponder serving the small station, and makes a frequency assignment to the small-station transmitter. This information is sent back to the small station, which sets its frequency synthesizer (3).

An alternative is to have a fixed-frequency small station that is told when it is to transmit its burst of information using time-division multiple access (TDMA). Many small stations may share the same part of the frequency band (typically 45 kHz in width) in the transponder. Since the exact frequency of each small station will be slightly different, the hub must be capable of rapidly accommodating this tolerance. The duration of a burst, for a high-resolution facsimile page, for example, could be long (e.g., one page per 500 ms). It could be very brief for credit card verification. Cofrequency microterminals would be queued up until there is an empty time slot. Since these services operate in nonreal time (like packet-switched data and, to some extent, freeze-frame television), the delay is acceptable.

A combination of variable-frequency microterminals (which frequency hop to an empty channel) and TDMA (to time-share the channels) is possible but an unnecessary complexity until congestion becomes important. The hub controls the actions of the microterminals just as an orchestra leader directs each player's timing, frequencies, and duration.

At the hub the network manager (generally, a minicomputer) is monitoring the network operation. If it detects a traffic growth that will lead to network overload, it may direct all (or some) stations to alter the modulation and/or multiple-access techniques (e.g., a pure ALOHA network may be reconfirmed to slotted ALOHA or TDMA during busy hours). This is done to reduce the likelihood of network blocking.

The network manager may also temporarily "busy-out" certain stations with low priority or less time-sensitive material. In these cases the channel requests would be deferred until the network load dropped.

Microterminal to Hub. In Figure 10.3 the originating station transmits (1) the encoded data (which include the destination address or addresses, the message length, and other information) to the hub station. At the hub the signal is received (7) and placed into a buffer storage device (10) as its destination address is read (18) and the number of errors counted (13). Once the address has been read, a request is made of the NOC (9) for connection to the destination. In some cases, there may be multiple destination addresses, and therefore the NOC must schedule multiple transmissions if the destinations use different frequencies, need different spot beams, and/or alert multiple stations within a transponder to receive the same time slot in the TDM transmission that will be coming.

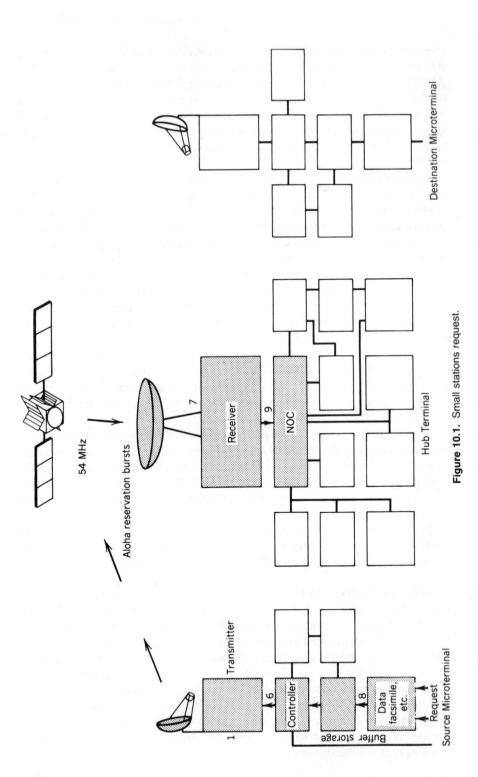

Figure 10.1. Small stations request.

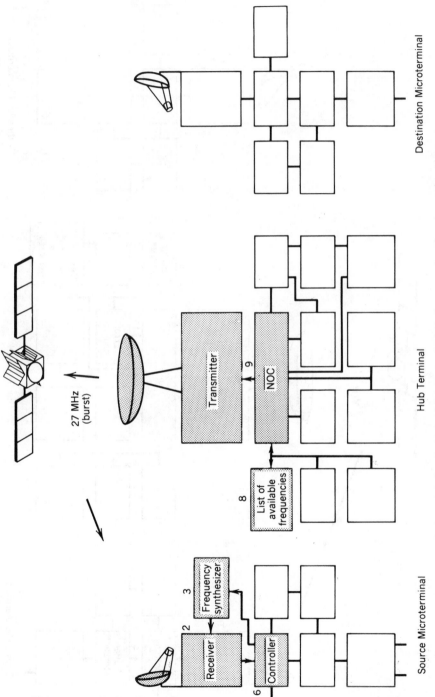

Figure 10.2. Network Operations Center recognizes the request and assigns a frequency.

Destination Microterminal

Hub Terminal

27 MHz (burst)

Transmitter

9 NOC

8 List of available frequencies

Source Microterminal

3 Frequency synthesizer

2 Receiver

6 Controller

107

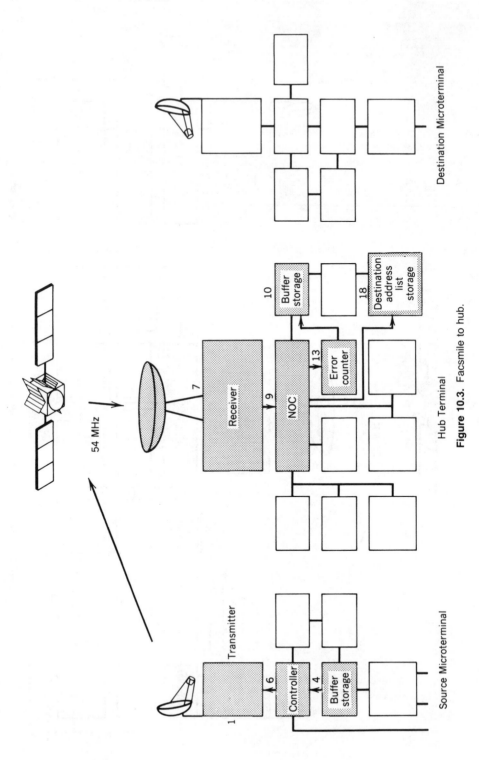

Figure 10.3. Facsmile to hub.

Source Microterminal

Transmitter

1

6 Controller

4 Buffer storage

54 MHz

Hub Terminal

Receiver

7

9 NOC

13 Error counter

10 Buffer storage

18 Destination address list storage

Destination Microterminal

108

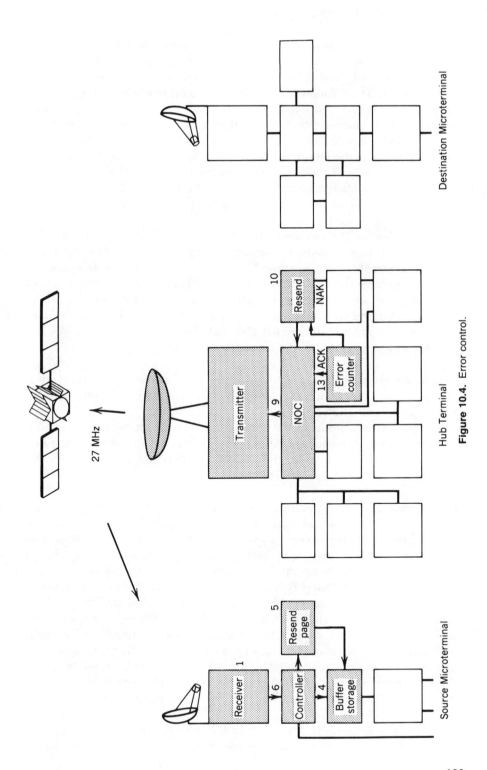

Figure 10.4. Error control.

Acknowledgment. In Figure 10.4 the NOC (9) transmits back to the originating small station (1) the message that it either accepts (acknowledgment or ACK) or does not accept (NACK) the transmission due to detected errors. At the originating station the data will be resent (5) if no acknowledgment or a NACK from the hub is received. No acknowledgment might occur if there was a heavy rain attenuation in either the uplink or downlink. If there is no response to the second transmission, it may be assumed that the link to the hub has troubles (either in the hub or due to hearing rain in the link). It makes little sense to continue facsimile transmissions until the link (or hub) is restored and the source's microterminals controller can be reset. If an ACK signal is received, the originating station prepares to send the next page.

Hub-to-Destination Microterminal. In Figure 10.5 the NOC assigns a time slot (11) to one or more destination stations in the downlink (CONUS) beam. The station sets its timing window (22) so that it can receive the transmission. In Figure 10.6 the buffer storage at the hub (10) is nondestructively read out and its contents for this page of data are transmitted to the destination station, which temporarily stores the information in a 200- to 500-kb buffer storage (24) while the errors are counted (20) and the destination station decides whether to accept or reject the transmission (23).

Destination Microterminal to Hub. In Figure 10.7 a successful reception is shown wherein the destination station sends (26) an acknowledgment (ACK) signal to the hub and releases its local buffer storage contents (24) into a printer (25), which produces the paper copies for the recipient or to a video display device. At the hub the ACK signal is received and after making sure that this is the last address for the page of data (14), the information for this page in the buffer storage (10) is deleted.

Error Handling. If the transmission to the destination had not been successful (e.g., due to heavy rain in either the uplink or downlink), either a NACK or no signal would be received by the hub and, depending on weather conditions at the hub, it may immediately reread the data from the hub buffer storage (10) and retransmit them (or it may wait): see Figure 10.8).

If the hub notices that it is receiving many NACKs (or no transmissions at all) from the small stations, it could conclude that the difficulty is local (either within the hub station or in the weather conditions between the hub and the satellite). If it notices that only certain stations are encountering difficulty, it may conclude that there is a local storm at their locations. It is at the system performance data (16) that these occurrences would be noted. With the pages safely in the buffer storage at the hub, they may be retransmitted until all the ACK signals are received (see Figure 10.9).

Administrative Matters. In Figure 10.10 the path of the final ACK signal is shown. The destination station (26) transmits its ACK to the hub, which,

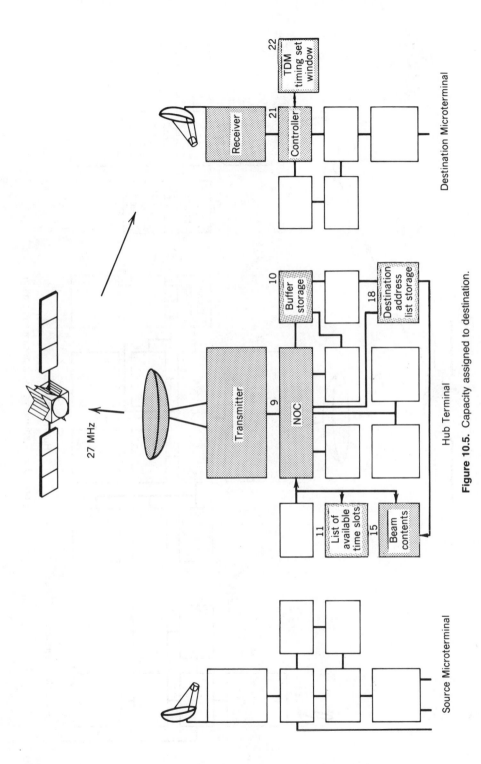

Figure 10.5. Capacity assigned to destination.

Source Microterminal

Hub Terminal

Destination Microterminal

27 MHz

22 TDM timing set window

21 Controller

Receiver

10 Buffer storage

9 NOC

Transmitter

18 Destination address list storage

11 List of available time slots

15 Beam contents

111

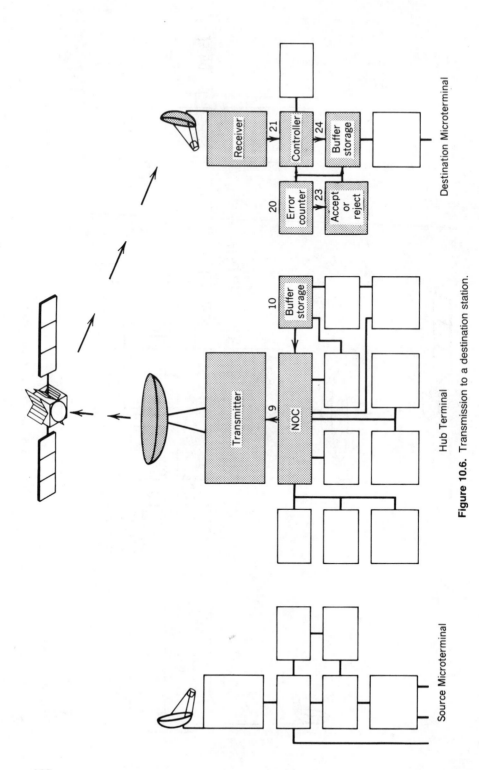

Figure 10.6. Transmission to a destination station.

Destination Microterminal

Receiver

21

Controller

24

Buffer
storage

20

Error
counter

23

Accept
or
reject

Hub Terminal

10

Buffer
storage

Transmitter

9

NOC

Source Microterminal

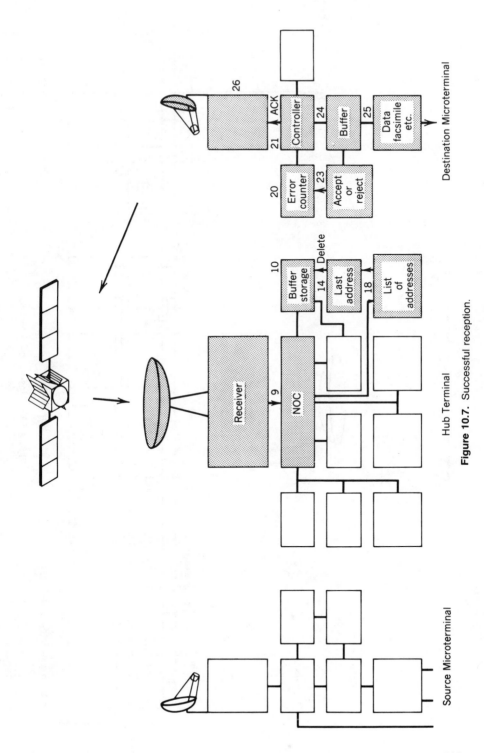

Figure 10.7. Successful reception.

Destination Microterminal

26

ACK
21 Controller
20 Error counter
23 Accept or reject
24 Buffer
25 Data facsimile etc.

Hub Terminal

Delete
10 Buffer storage
14 Last address
18 List of addresses

Receiver
9 NOC

Source Microterminal

113

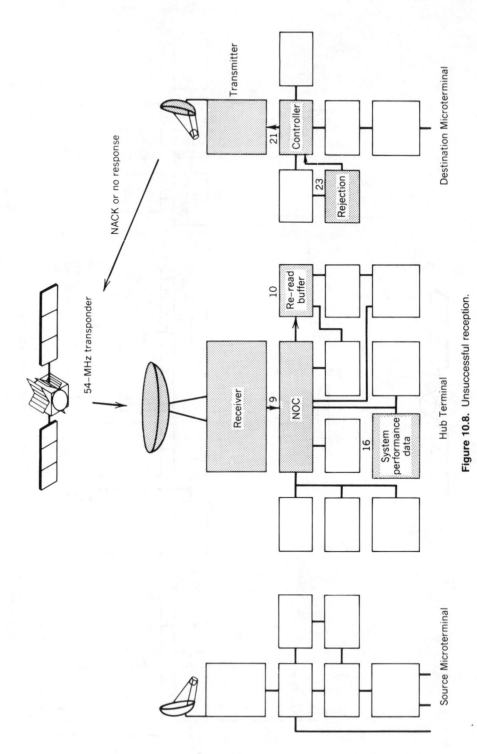

Figure 10.8. Unsuccessful reception.

NACK or no response

54-MHz transponder

Transmitter

Controller

21

23 Rejection

Destination Microterminal

Receiver

9 NOC

10 Re-read buffer

16 System performance data

Hub Terminal

Source Microterminal

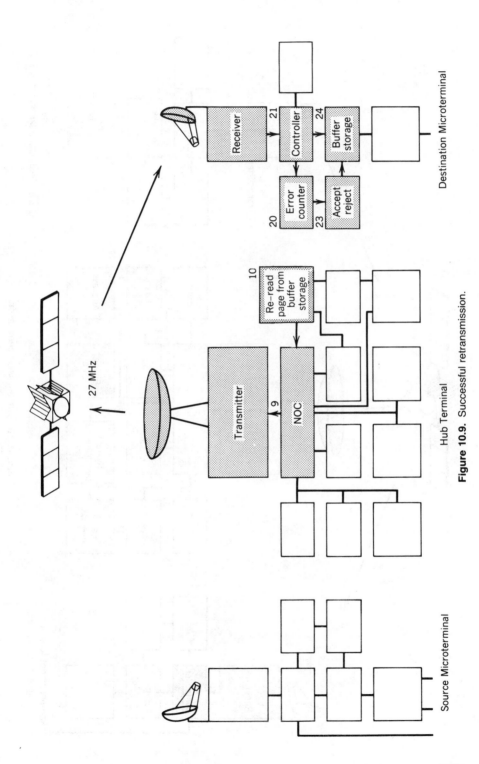

27 MHz

Destination Microterminal

Receiver

Controller

Buffer storage

21

24

Error counter

Accept reject

20

23

Re–read page from buffer storage

10

Transmitter

NOC

9

Hub Terminal

Source Microterminal

Figure 10.9. Successful retransmission.

115

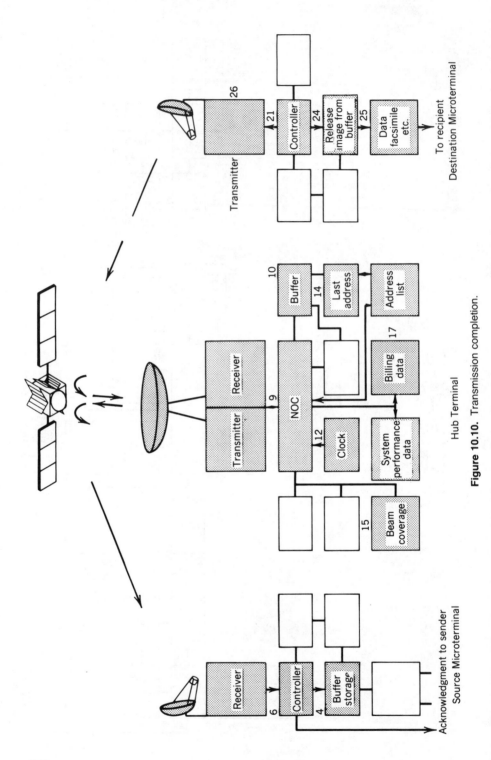

Figure 10.10. Transmission completion.

after consulting its local clock (12), prepares the billing data (17) with the time of delivery, deletes the page for the buffer storage (10) (if it is a last address for that page, 14), and finally, signals the originating station controller (6) to tell the originator that all the pages were delivered successfully.

In general, multiaddress data and facsimile would all be delivered simultaneously unless there was a rain outage at one of the destination microterminals which would require a later retransmission. If the destination microterminals are limited in the number of transponders that they can receive, multiple transmissions of a page (on different transponders) may be necessary. In a system using a satellite with multiple beams, the NOC must also know which beams cover which destinations (15).

Since the NOC knows which receiving stations are busy with other traffic, it may be necessary to keep some pages in storage (10) until the destination station is free to receive these pages.

10.1.2 Credit Verification

Microterminal to Hub. When a department store clerk is handed a credit card, he or she places it into a scanning device that reads the card number into a telecommunications buffer elsewhere in the shopping center. The value of the sale is also entered. The buffer determines the card type and either places a telephone call or routes the inquiry to the microterminal controller. The controller adds the merchant's address information, and so on, and prepares the microterminal for transmission.

In some systems the transmission is made on an ALOHA contention system. If the message reaches the hub (without colliding with another user) an acknowledgment is returned from the hub to the microterminal (the microterminal may not be able to see its own uplink transmission due to its power level). If no acknowledgment is received, the microterminal waits for a random time and retransmits (hoping to avoid another station's transmissions). As long as traffic is light, ALOHA works fine. When it gets busy, the system can suddenly block itself and exhibit abrupt bistable time delays. See Section 13.8 for more information on ALOHA.

Hub Activities. The hub looks up the card number. Based on the credit balance, history, and the amount of the sale, the hub computer provides a "yes," "no," or other indication. "Other" may include such coded messages as "approved, but have the customer call American Express," "the card has been stolen," "the account has been closed," or "do not give the customer the card back." A series of digits (the "authorization number") may be added. For more information, consult Section 14.6.4.

Hub to Microterminal. The hub combines the return address (see above) and the "yes," "no," or "other" response. This message, together with others, are transmitted sequentially (by TDM), from the hub. The microterminals listen to all messages from the hub on their frequency (just like a taxi driver monitors his radio for assignments). When its address is heard, the message is accepted and sent via its controller to the originating clerk.

Error Handling. In the simplest system there is no control over rain outages, ALOHA collisions, and so on. The clerk may rescan the card until there is an answer. As in all fund transfers, forward error control is needed to prevent costly errors. Many code forms are available using parity and check sum bits. These extra bits increase the amount of traffic.

10.2 HUB CONTROLLER

The hub contains a network operations center (NOC) controller that monitors all networks being operated via the hub. These networks may be for several clients or for departments of a large organization. There may also be subnetworks arranged on an administrative, geographical, traffic type, or data-rate basis. Video teleconferencing is one subnetwork example.

A hub may be shared by many different corporations and organizations. The networks may be reconfigured as the type of traffic changes during the day or week.

The networks and subnetworks may each have their own dedicated space segment (power, bandwidth, and transponders), have a reserve pool of space segment (for peak busy hours), or dynamically share a common resource.

A hub may operate with several satellites (each via its own hub earth station antenna), but in general, each microterminal can only use one satellite. In at least one instance, Niagara Mohawk Power, there are two microterminals at some sites each aimed at a different satellite for full-path redundancy. Of the 720 locations in their network, 180 have dual stations.

The hub controller uses one or more computers, one or more PBXs, and contains the channel interfaces for intranetwork and internetwork (to other networks, often via the telephone system) connections. Packet-switching equipment is used in many hubs. Conversions from one format (SCPC, TDMA, ALOHA, etc.) to another (frequently TDM) are made at the hub.

10.2.1 Privacy

The privacy of commercial telecommunications is of worldwide concern. This is especially true with satellite communications, where it is easy for a third party to intercept transmissions.

The ability to receive information being broadcast or sent to other parties is an element of concern in satellite telecommunications.

Since the casual listener can generally monitor only one channel at a time, there is a limited amount of information available from two-way services. Even if the codes could be read, each eavesdropper only knows the question and has no access to the answer (or receives the answer and does not know the question).

For services that broadcast information continually (point to multipoint) the acquisition of data bases (e.g., the stock market transactions) might be valuable to an evesdropper.

In most of these instances, sufficient scrambling or higher-order modulation techniques (including error correcting) are used. The objective is to make unauthorized interception too expensive for most earth station owners that do not have the complex decoding equipment.

In the case of spread-spectrum systems, the despreading code must be known and a lock achieved between the transmitter and receiver. Since locking the receiver to the transmission takes a finite amount of time, the likelihood of accidentally discovering the right despreading combination is substantially lessened. Since the code is constantly changing, the odds of deriving useful information is even smaller. If multiple layers of encryption are used, an eavesdropper may never be able to tell even when the code for the first layer has been cracked.

In the case of TDMA, TDM, and frequency-hopping SCPC, one channel contains an almost random sequence of data to (or from) many hundreds of microterminals. The listener can gain little or no information. The hub may need to transmit random bit sequences between real messages, to reduce interference to other users. These random sequences will further confuse unauthorized listeners.

The microterminal transmissions are much more secure than the terrestrial microwave or copper telecommunications. Additional levels of security are available commercially using a DES method password access and other commercial methods.

10.2.2 Billing

Chapter 11 discusses the various protocols and standards. Inherent in these protocols is an indication of the source and sink of the telecommunications. Since all of this information must pass through the hub station, the hub may decode and "read" these addresses, determine the destination(s) and duration of each message, and prepare monthly billings for the using organization(s). In some instances, the network will bill all of the expenses to a general expense fund. In most instances it will be billed back to the individual organization, in some cases to the individual users, and in other cases to a specific shop order. The protocols provide sufficient flexibility for this type of charging. Individual users may pay on the basis of the amount of information transmitted.

10.3 HUB SHARING

A typical hub may cost $1 million to construct by the time the RF, data processing, civil works, ground, and so on, are included. As shown in Table 4.1, operation of a hub is even more expensive. There are few reasons why multiple users cannot share the hub and the personnel. For safety reasons, the number of people has been set at two per shift. The administration changes are insensitive to traffic; thus, adding more networks (that use the same type of equipment) may not call for additional people or expenses.

Hub sharing permits new networks to be established quickly and economically with minimum financial investment and no technical involvement (the original operator provides the hub managment function). For manufacturers it is a way to sell equipment initially and then retain the customer by demanding that all new microterminals meet its hub standards.

10.4 SINGLE OR MULTIPLE HUBS

Most systems today use a single hub. It is expected that in future years some of these will grow to have multiple hubs (thereby providing additional redundancy against the possibility of a rain outage at the principal hub). The second hub may be located at an existing teleport or shared with another network that uses the same type of equipment at a separate hub. A mutual restoration agreement is possible between the two independently owned hubs. The secondary hub also provides redundancy in the case of heavy rainfall, fire, flood, earthquake, interference, and so on. It should be located far enough away to be in a diverse climate. Preferably, one should be in a low-rainfall area (e.g., in the southwestern U.S. desert) to minimize rain outages.

10.5 SPACE SEGMENT SOURCE

The network operator may:

1. Use its own space segment (leased or purchased from a satellite common carrier)
2. Share space segment with other users
3. Use the space segment that may be supplied by the satellite carrier as part of an overall package, which includes the microterminals and the hub

The latter case is exemplified by the C-band service offered by AT&T under the name "Skynet." These options are discussed in Section 9.2.

10.6 RAIN CONDITIONS

Precipitation can reduce the signal level between both an earth station and its satellite and the satellite and its earth station. In general, snow is not a problem (unless it collects on the reflecting surface of the antenna), but rainfall is a matter for concern. It is the *rain rate* (in mm/hour), *not* the total annual amount of rain (in mm), that is important [1,2]. The rain rate is the amount of rain at any instant within the beam between the earth station and the satellite.

Rain has three effects. The first is to increase the attenuation (through the absorption of the signal). The second is to cause a rotation of the linear polarizations which are used in the United States for transmission. Because the raindrop is asymmetric, it is not surprising that the raindrop will attenuate more of one polarization (along the raindrop's length) than the other (along the width). This causes a rotation of the polarization of the two signals. Consequently, polarization isolation degrades under these circumstances.

The third effect is the increase in the sky noise temperature as seen by a receiving antenna at 12 GHz. In an extremely heavy rainstorm, the earth station antenna will be looking into a large volume of rainfall (which is at near-ambient temperature). Instead of the earth station looking into cold space (at a few kelvin), it is now looking at a warm noise source. A reasonable value in rain is 50–200 K and is related to the attenuation value.

This situation is not true in the case of the satellite antenna. Since the satellite antenna looks at a large area (e.g., the entire United States), it will integrate the temperature of rain with the nonrain areas. As a matter of fact, rain clouds have tops (as seen by the satellites) which are much cooler than the underlying earth, and therefore the rain actually reduces the noise temperature, but only very slightly, as indicated above.

If possible, the selection of the hub station should be done on the basis of the rainfall statistics. Figure 10.11 shows the location of present hubs superimposed on a map showing the N.A.S.A. rain regions within the United States. Some locations are obviously better than others at K_u-band. This will be even truer when the K_a-band (30/20 GHz) services are initiated. When it is possible to select a hub location based on optimum transmission, a site in the U.S. Southwest (e.g., Phoenix, Arizona) is preferable. In many cases, other considerations take priority. This becomes true when the hub must be co-located with existing computer or administrative functions.

For comparison, Figure 10.12 shows a map of Europe and the CCIR regions [3]. In general, the rain rate is higher in the United States than in Europe. Table 10.1 shows the rain rates in North America and Europe.

Figure 10.13 shows the typical U.S. rain attenuation at 12 GHz for various availability or outage criteria. Most systems are designed to operate between 99.75 (typical terrestrial network performance) and 99.95% of the time. In New York (rain region D_2) a margin of 2–5 dB is needed to achieve these

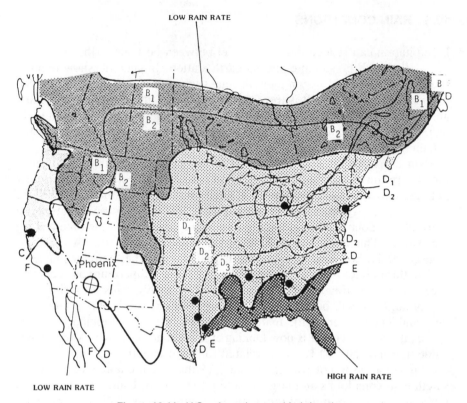

Figure 10.11. U.S. rain regions and hub locations.

levels in a typical year. Figure 10.13 cannot be used with Figure 10.12 because of the different rain rate criteria.

10.7 SUN OUTAGES

When the sun moves behind a satellite, the antenna "sees" a hot object and the noise temperature climbs. In conventional service with narrow-beam (large physical diameter) antennas, the sun can occupy an appreciable portion of the beamwidth of the antenna. Since the sun is approximately 25,000 K (it varies as a function of the solar/lunar cycle and the radio frequency), it may raise the system noise temperature to the point where the carrier-to-noise ratio drops below threshold.

The timing of a sun outage can be predicted years in advance, so that traffic may be delayed or rerouted (where possible) during any brief sun outage. These outages occur in two seasons of the year (spring and fall) for several days. If coverage during these periods is critical, two satellites (with

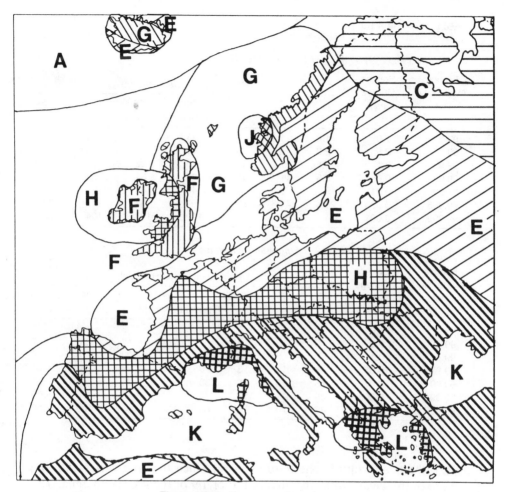

Figure 10.12. European rain regions.

two microterminals) should be used. A dial-up telephone line may be a more economical alternative during these brief occurrences.

In the case of a typical spread-spectrum service (which is engineered to tolerate large amounts of noise from adjacent satellite interference and terrestrial microwave services), the use of spread spectrum and the very wide beamwidth of the small C-band stations results in a small change in noise temperature. This is a subtle advantage of the small antenna spread-spectrum systems, due to both the processing gain and the antenna.

If the sun subtends about 0.5° (0.2 square degrees) and a 60-cm (2-ft) antenna has a 4-GHz beamwidth of 8.75° (or 60 square degrees), the sun occupies 1/300 of the beam and the sun's noise temperature is diluted substantially. If the same antenna is used at three times the frequency (= 12

TABLE 10.1   Rain Rates

North American Rain Rates

Service Availability (%)	Service Outage (%)	Rain Region (Crane Model)						
		B	C	D1	D2	D3	E	F
				(Rain rate, mm/h)				
99.0	1.0	1.8	1.9	2.2	3.0	4.0	4.0	4.0
99.5	0.5	2.7	2.8	4.0	5.2	7.0	8.5	2.4
99.9	0.1	6.8	7.2	11	15	22	35	5.5
99.95	0.05	9.5	11	16	22	31	52	8.0
99.99	0.01	19	28	37	49	63	98	23
99.995	0.005	26	41	50	64	81	117	34
99.999	0.001	54	80	90	102	127	164	66

European Rain Rates

Service Availability (%)	Service Outage (%)	Rain Region (CCIR Model)						
		C	E	F	G	H	J	K	L
					(Rain rate, mm/h)				
99.0	1.0	2	1	2	3	2	8	2	2
99.7	0.3	3	3	4	7	4	13	6	7
99.9	0.1	5	6	8	12	10	20	12	15
99.97	0.03	9	12	15	20	18	28	23	33
99.99	0.01	15	22	28	30	32	35	42	60
99.997	0.003	26	41	54	45	55	45	70	105
99.999	0.001	42	70	78	65	83	55	100	150

GHz), the beamwidth is one-third as wide. Since the area varies with the square of the beamwidth, the area is $1/3^2$ or 1/9 of the 4-GHz antenna. The sun occupies 9 times as much relative area in the beam and thus has a more significant effect (see Figure 10.14) at 12 GHz. Since the beam is narrower, the sun moves through the beam in one-third the time.

10.8   INTERFERENCE

Spread-spectrum multiple-access systems can tolerate higher levels of noise and interference than TDM or SCPC, due to the method used of transmitting and receiving the signal. In the case of the SCPC and TDM, systems typically require carrier-to-interference (C/I) ratios of 15 dB or more. Through the use of forward error correction and other techniques, some improvement may be obtained but at the price of additional bandwidth and bit rate. As the C/N is lowered (e.g., by use of smaller antennas), the C/I becomes more critical.

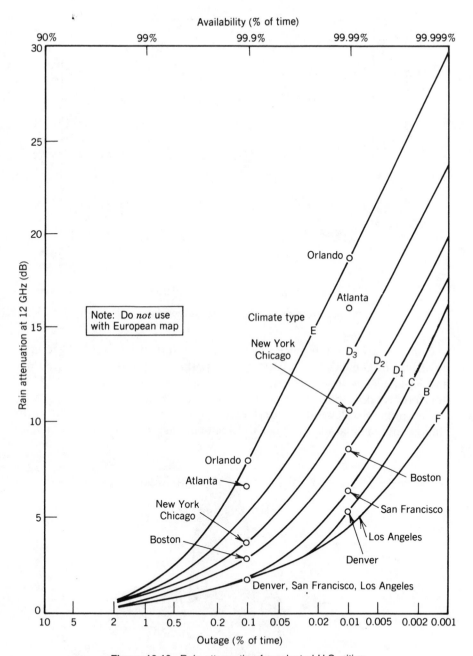

Figure 10.13. Rain attenuation for selected U.S. cities.

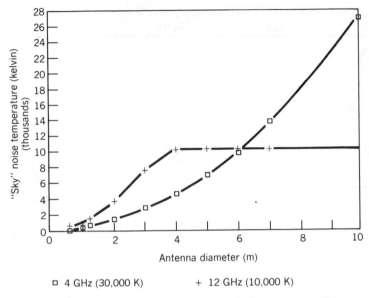

Figure 10.14. Noise temperature during sun transit (antenna pointed directly at sun).

Concerns over interference (particular from terrestrial sources) are very important in C-band, but relatively unimportant at the domestic U.S. K_u- and K_a-bands. A few terrestrial K-band transmitters remain. Most of the interference at K_u-band is to or from adjacent satellites.

The principal element in interference control is the design of the earth station antenna and in particular its sidelobe content along the orbit plane.

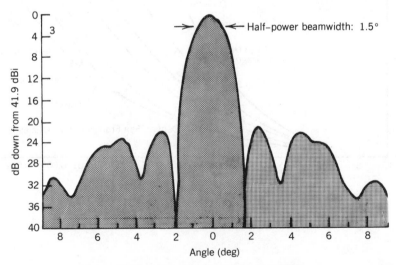

Figure 10.15. 1.2-m antenna pattern at 12 GHz.

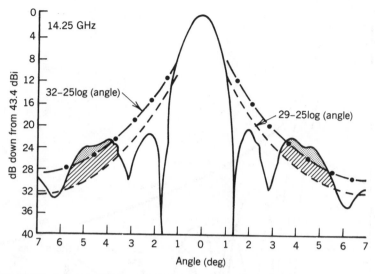

Figure 10.16. Antenna of Figure 10.15 at 14.25 GHz with CCIR and FCC requirements.

Figure 10.15 shows a typical response for a 1.2-m antenna used for reception at 12 GHz. If this antenna was used for transmitting, its pattern would be as shown in Figure 10.16. Figure 10.17 shows the maximum permitted antenna gain. The CCIR limit is presently 3 dB higher [$32 - 25 \log_{10}(\theta)$]. Theta is the off-axis angle in degrees. Figure 10.18 shows the measured pattern for an antenna that meets the FCC requirements.

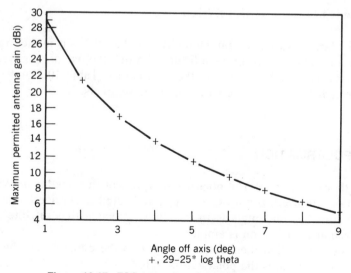

Figure 10.17. FCC transmit performance requirements.

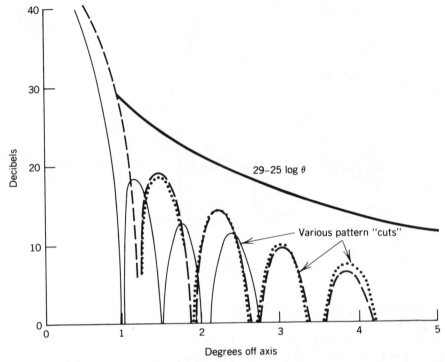

Figure 10.18. Acceptable 14-GHz pattern.

10.9   SATELLITE POWER LEVELS

One of the largest variances among the K_u-band satellites is in their radiated power levels. This is further complicated through the use of spot, regional, and national (CONUS) beams on these satellites. Table 10.2 and Figures 10.19 through 10.26 show the variation in these levels.

10.10   POLARIZATION

Finally, there is the matter of polarizations. Linear (horizontal and vertical) polarization are used in both C and K_u fixed satellite service bands in the United States and by Intelsat. In the direct broadcasting satellite service band, circular polarization is employed.

Most satellites align their polarizations with the equator and the line of the poles. RCA rotates the polarization 5° for their Satcom K series. In the

TABLE 10.2 EIRP on Various Cities[a]

Case	Satellite	Beam	N.Y.	Chi.	Atl.	Mia.	Dal.–Ft W.	Den.	L.A.	S.F.–O.	Seattle
1	Anik-C	E + EC	32	33							
2	Anik-C	W + WC						30		25	33
3	ASC-1	CONUS	44	43	43	38	42	43	43	44	40
4	Expresstar	Spot									
5	Expresstar	Regional	53.5	50	50		41	51	47	51	54
6	Gstar	CONUS	43.5	44	44	41	43	41	42	42	40
7	Gstar	W region		40			42	44	43	45	43
8	Gstar	E region	46	45.5	46.5	42.5					
9	Gstar	E spot	47	47	41		41				
10	Gstar	W spot							48	50	48
11	Satcom K	CONUS	45	45	46	39	44	44	45	45	43
12	Satcom K	E region	47	47	48	41	47				
13	Satcom K	W region						49	48	48	46
14	SBS 1–4	CONUS	47	45	46	43	44	43	43	43	43
15	SBS-4	E spot	52.5								
16	Spacenet	CONUS	44.3	43.2	43.7	38	42.9	43.8	44	43.8	40.5
17	Spacenet III	E spot	46.4	44	45	41	42				
18	Spacenet III	W spot						46	47	47	46

[a]In dBW.

129

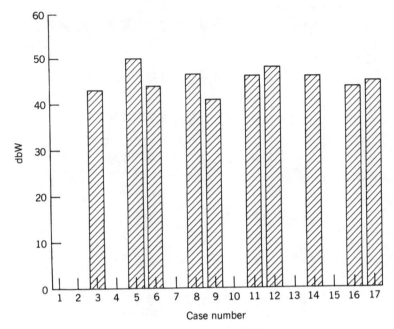

Figure 10.19. Atlanta EIRP.

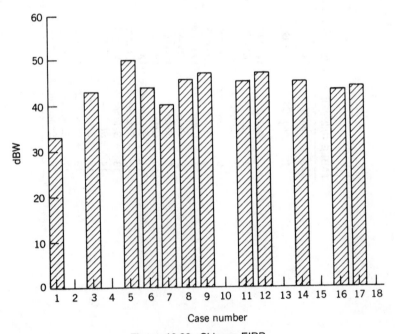

Figure 10.20. Chicago EIRP.

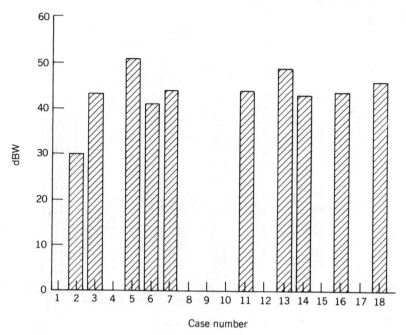

Figure 10.21. Denver EIRP.

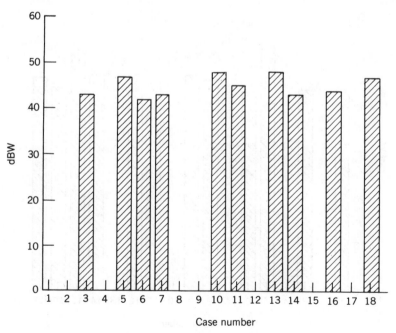

Figure 10.22. Los Angeles EIRP.

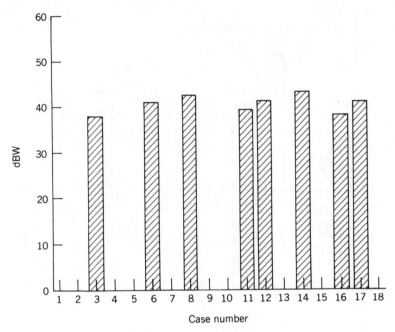

Figure 10.23. Miami EIRP.

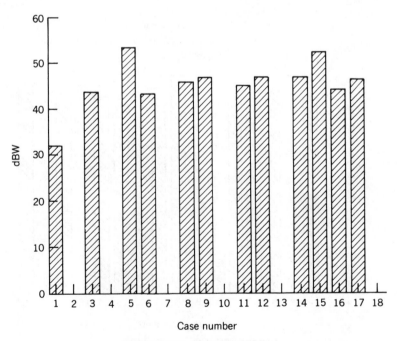

Figure 10.24. New York EIRP.

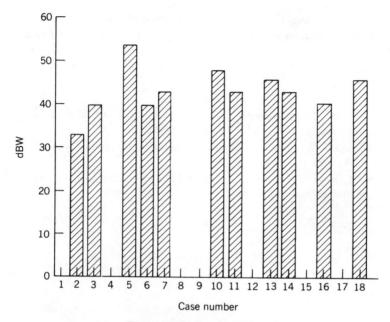

Figure 10.25. Seattle EIRP.

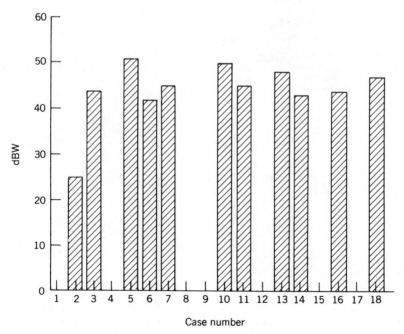

Figure 10.26. San Francisco EIRP.

case of Gstar, the angle is rotated 26°. The senses of the uplink and downlink polarizations are not entirely consistent along the orbit arc (see Section 12.4).

REFERENCES

1. L. J. Ippolito, Jr., *Radiowave Propagation in Satellite Communications.* Van Nostrand Reinhold Co., Inc., New York, 1986.
2. T. Pratt and C. W. Bostian, *Satellite Communications.* John Wiley & Sons, Inc., New York, 1986.
3. Anon. CCIR Volume V, *Propagation in a Non Ionized Medium,* International Telecommunications Union, Geneva, Switzerland, 1987.

11

STANDARDS

11.1 INTERNATIONAL STANDARDS

The CCITT (Committee Consultative Internationale de Télégraphique et Té-
léphonique) is the international agency primarily concerned with the telephone and data systems. The CCITT's task is to make technical recom-
mendations about telephone, telegraph, and data communications interfaces.
Many of these become widespread industry standards: for example, V.24
(EIA-RS-232) and X.25, which delineates the interface between a computer
and a computer network.

Many of the microterminal networks use the established protocols of the
CCITT as well as the International Standards Organization (ISO). In the
United States, the Electronics Industry Association (EIA) standards are also
widely used: for example, EIA-RS-232, which defines the placement and
meaning of the pins on the connector used by most terminals.

11.2 PROTOCOLS IN USE

11.2.1 Layers

The Open Systems Interconnection (OSI) reference model breaks the process
of computer-to-computer communications into a set of seven layers. The
top layer (7) is the application layer, (6) the presentation layer, (5) the session
layer, (4) the transport layer, (3) the network layer, (2) the data link layer,
and (1) the physical layer.

The top three layers (7 to 5, or applications to session) form a group of

TABLE 11.1 Network Layer Nomenclature Compared to ISO Reference Model

Sopho-Net	Arpanet	Layer	ISO	Decnet	SNA
Information domain	User	7	Application	Application	End user
	FTP–Telnet	6	Presentation		NAU services
		5	Session		Data flow control
Communication domain	Host-to-host	4	Transport	Network services	Transmission control
Transfer domain	Source–dest.–imp.	3	Network	Transport	Path control
Link domain	Imp.–imp.	2	Data link	Data link control	
Physical domain	Physical	1	Physical	Physical	Physical

layers where most of the functions for manipulation of information take place. These functions take place irrespective of the physical medium for transmission of data being used. The bottom three layers—data link (3), network (2), and physical (1)—contain the components used to transmit messages. Layer 4, the transport layer, is the interface between the end users and the communications network. Table 11.1 shows the network layer nomenclature compared to that of the ISO reference model.

11.2.2 Arpanet and Satnet

Not all of the 1986 U.S. computer communications networks conform to this nomenclature or functions. Existing networks (such as Arpanet, SNA, and Decnet*) use variations of the ISO scheme.

Arpanet, created by the Defense Advanced Research Projects Agency of the U.S. Department of Defense, connects hundreds of computers and spans over half the earth, from Hawaii to Norway, England, and Italy. Part of Arpanet is Satnet, a packet-switched satellite communications net operating on SCPC channels on an Intelsat transponder. Satnet uses a distributed packet-switching system operating with the Aloha access protocol in the FPODA (Fixed Priority of Demand Access) or CPODA (Contention Priority of Demand Access) schemes. The network uses two 64-kb/s SCPC channels

---

*These terms are defined in later sections of this chapter.

for all of the traffic between five nodes. Arpanet uses the TCP transport protocol. It has remote-log-in facilities and uses FTP for file transfer.

Remote microcomputer terminals can access the Arpanet through dial-up lines. These terminals can use the Kermit protocol to transfer files within the net. Kermit was developed by Columbia University and is supported by many mainframe machines.

Satnet is a distributed communications network. This means that there is no master control station, but each node is constantly updated with network timing and traffic data. The Satnet protocols require that each node located at the Intelsat earth station can receive its own transmissions. This, in turn, requires the use of global beams, or at least beams that simultaneously cover the area of interest of the network. This presented no problem for the Intelsat C-band global beams and 30-m stations used by Satnet. Since, in most microterminal networks operating in K_u-band, the small earth stations cannot receive their own transmissions, a distributed network architecture such as Satnet cannot be used without some modification.

11.2.3 IBM's SNA

SNA (System Network Architecture) is used extensively by IBM and its customers. An SNA network connects a set of nodes of four types, including terminals, antennas, front-end processors, and host computers. Each node contains one or more network addressable units (NAUs) which have network addresses. A NAU is the software that facilitates network access by outside processes.

The microterminal uses the 3725 cluster controller protocols. At the hub, a 3270 emulator is used. Satellite-based networks connecting customers that use IBM mainframes will probably continue to use the SNA network architecture.

11.2.4 Decnet

Decnet, which is a set of programs and protocol produced by Digital Equipment Corporation, uses an architecture called DNA (Digital Network Architecture). Decnet allows customers to set up private networks. Decnet architecture has five layers, with the lowest two, physical and transport layers, corresponding to the physical and transport layers of ISO. Decnet has no sessions layer. Its application layer combines the functions of ISO's application and session layers.

11.2.5 X.25

X.25 is an international standard defining, via X.21 and X.21 bis, three layers of communications (the physical layer, the frame layer, and the packet layer).

X.25 is widely used inside and outside the United States and will certainly be used in the United States by networks that have international connectivity. In some cases the X.25 packet assembler/disassembler is integrated into the microterminal controller. Packet switching has the advantage of requiring only occasional satellite capacity. Most of the time the channel is idle, thus available to other users. Various demand assignment methods maintain the use of the space segment resource.

11.3   A PACKET-SWITCHED NETWORK EXAMPLE

Sopho-Net (TM), of Philips Export B.V., is a packet-switched wide-area network designed to fulfill the requirements of business communication systems. The main characteristics of Sopho-Net are:

1. The ability to incorporate computers and terminals of different origins, types, and sizes in the same network
2. The ability to use all existing and foreseeable transmission resources and switching systems, such as private lines, public switched telephone networks, PABXs (private automatic branch exchanges), X.21 and X.25 networks, local area networks, satellites, and so on
3. The facility to access non-Sopho-Net networks via gateways
4. The utilization of a powerful routing method which compensates for resource failure and adapts itself to traffic loads
5. A comprehensive end-to-end protocol providing error recovery and flow control
6. A unique approach to the problem of protocol conversion
7. The distribution of the network management and maintenance function, resulting in high availability and extensibility of the network
8. The flexibility and modularity of the system, due to the multi-micro architecture of the supporting hardware
9. A wide range of system sizes in terms of traffic, number of connections, nodes, and so on, due to the use of advanced technology in the hardware design

Of interest to this study is the second point above: the ability of Sopho-Net to work with communication satellite links. The architecture and organization of Sopho-Net is such that it can readily connect to microterminal links. Arpanet already has a satellite link connectivity, albeit using Intelsat Standard-A earth stations with their 30-m antennas.

Thus Arpanet, through its Satnet subnet, serves as a precedent for including satellite links into sophisticated computer-to-computer communications networks. It is expected that Sopho-Net will include microterminal links where these are found to be economically desirable.

Sopho-Net has a logical structure consisting of layers which are called domains. It has three fundamental domains: the information domain; the communications domain; and the transfer domain. Each domain implements two functions: data exchange and network management. These functions encompass the essential interface and protocol conversions and network routing and transmission resource management in the network.

11.4 U.S. DIGITAL STANDARDS

Table 11.2 shows the basic digital transmission rates in North America and Europe. The U.S. transmission rates (e.g., T1) are also referred to as D or DS (for digital system) rates. If fiber optics is used, the prefix F may be used (e.g., FT3). The voice frequency (VF) capacity may be increased by using

TABLE 11.2 Standard Digital Transmission Rates

System	Rate (Mb/s)	VF Channels[a]	System Facility	Error Rate	Notes
T0	64 kb/s	1	Twisted pair or satellite	1×10^{-6}	Digital voice or data
T1	1.544	24	Paired cable or satellite	1×10^{-6}	Basic North American system
T1C[b]	3.152	48	Paired cable or satellite	1×10^{-6}	Expansion system for existing T1
T2[b]	6.312	96	Low-capacitance paired cable	1×10^{-7}	
T3	44.736	672	Radio		
T4	274.176	4032	Fiber or radio	1×10^{-6}	High-density long-haul system
T5	560.160	8064	Fiber	4×10^{-7}	Typical fiber system
E1	2.048	30	Paired cable, radio, or satellite	10^{-10} per km	Basic European system (CEPT)
E2	8.448	120	Radio or satellite	10^{-10} per km	Short- and medium-haul trunks
E3	34.368	480	Coax or radio	10^{-10} per km	Short- and medium-haul trunks
E4	139.264	1920	Coax, fiber, or radio	10^{-10} per km	Intercity trunks
E5	565.148	7680	Fiber	10^{-10} per km	Typical fiber system

[a]One-way voice-frequency channels at 64 kb/s.
[b]Rarely used.

TABLE 11.3 Composition of the Lower-Level Digital Systems

	American T1		European CEPT E1
VF channels	24	24	30
Signaling channel	0	0	1
Framing channel	0	0	1
Total channels	24	24	32
Number of bits/channel	7	7 5/6	8
Signaling bits	1	1/6	0
Total bits/channel	8	8	8
Total bits per sampling (frame)	192	192	256
Frame bit	1	1	0
Bits per frame	193	193	256
Sample rate (kHz)	× 8	× 8	× 8
Total line rate			
Kb/s	1544	1544	2048
Mb/s	1.544	1.544	2.048

32 kb/s instead of 64 kb/s per voice channel. Table 11.3 shows the details of the T1 and E1 (European) channels.

11.5 RFP PROCESS

When a company issues a request for proposal (RFP), it usually describes the microterminal network that it has conceived as fulfilling its requirements. These needs are based on internal studies (often assisted by an outside consultant). Most of these networks are for use within a corporation only. Compatibility with outside networks (e.g., the telephone companies) is not always mandatory. More important than outside compatibility is the ease of connecting existing equipment within the company. This includes word processors, computers, computer-aided design/computer-aided manufacturing (CAD/CAM), and the internal voice networks. Teleconferencing may or may not be included as a requirement. The manufacturers prepare their bids based on the RFP but are free to deviate (at the risk of being rejected) from the specifications. When this is done, it is usually to incorporate an existing product (or planned product) into the bid, thereby lowering the overall price. Since price is an easier element to judge than a promise of performance, reducing the price is often important.

As a result of the RFP process and the open bidding, either the user or the manufacturer may impose their own concept of "standards." This is particular true when it comes to transmission modulation methods (bit rates, forward error correction, etc.).

In some instances, deviation from standards may be deliberate. If the manufacturer also provides and operates the hub, it may be able to keep the network users from using equipment that is "nonstandard" (competitive) to the network.

12

REGULATIONS

12.1 JURISDICTIONS

12.1.1 International Jurisdiction

The International Telecommunications Union (ITU) and its International Frequency Registration Board (IFRB) assigns and administers telecommunications frequencies internationally. In general, the United States adheres to the Region Two Frequency Tables. There are some minor variations within the United States and, in particular, in the assignment of some short-range terrestrial services.

12.1.2 Federal Jurisdiction

The three principal federal agencies that oversee microterminals are the Federal Communications Commission (FCC), the National Telecommunication and Information Administration (NTIA), and the Occupational Safety and Health Administration (OSHA).

The Federal Communications Commission is charged with the responsibility of administering the radio-frequency spectrum for nongovernment uses. This includes the licensing of stations, satellites, and operators.

NTIA is the central agency for government users. The government frequency assignments are, in general, different from those used by the civil sector. Since commercial operators are allowed to bid on government contracts dealing with telecommunications, there are some government applications of commercial frequencies, but this is done on a contract basis and is no different than making arrangements for long-distance telephone services from one of the common carriers. OSHA is concerned with the health aspects and, in particular, with the electromagnetic radiation concerns.

12.1.3 State Jurisdiction

Each of the U.S. states has a public utility commission (sometimes referred to as the "public service commission" or "corporation commission") that oversees *intra*state telecommunications pricing and services.

If any part of a service crosses a state line, it becomes an *inter*state service and is therefore subject to the regulation of the Federal Communications Commission. In most cases, the federal regulations are less stringent for both the user and the supplier and therefore are preferred. Most microterminal networks will cross state lines (sometimes deliberately), and therefore state regulation of satellite networks is virtually nonexistent. Since commerce between the states is totally transparent and the coverage of the satellite antenna patterns is multistate, it seems unlikely that most individual states will get involved in microterminal regulations. The one exception to this situation is Alaska, which due to its size and geographic isolation from the other 49 states, regulates the primary satellite common carrier. Alascom is also the terrestrial telecommunications carrier for the state and therefore is the dominant carrier.

Most of the K_u-band satellites cover only the 48 contiguous United States (CONUS). At C-band, most satellites cover all 50 states (unless limited by the orbital location).

12.1.4 Local Jurisdiction

Local regulations at the county, city, town, or district basis may impose regulations on the locations of earth stations for zoning and aesthetic reasons.

On January 14, 1986, the FCC (Docket Case Report DC-362 in CC 85-87) held that some local ordinances discriminated against satellite reception antennas compared to other forms of antennas. This has limited the powers of the local jurisdictions.

The local authorities (states and municipalities) can enact ordinances restricting these antennas if:

1. "A reasonable and clearly defined health, safety, or aesthetic objective is stated."
2. They do not "impose unreasonable limitations on, or prevent the reception of satellite-delivered signals" or impose "excessive costs" on the users.
3. It does not discriminate on the basis of antenna size or shape.

The ruling goes on to state that the municipalities cannot "prevent satellite reception under any circumstances." If a municipality limits other types of antennas (amateur radio, microwave TV, and in some instances regular FM and TV antennas) to preserve the "historic character" of a community, it may also do so with satellite antennas.

The FCC feels that U.S. citizens have a "federal right to construct and use antennas to receive satellite-delivered signals" that should not be limited by the states or local municipalities. It is expected that this will leave the local jurisdictions with a choice as to how, or even if, to regulate transmit and receive/transmit stations in the microterminal class.

Challenges on antenna size, shape, and aesthetics may be impossible. This leaves the issue of "a reasonable and clearly defined health or safety matter." Although OSHA has defined an electromagnetic radiation safety level, some states have been considering lower levels. In most microterminals, the radiation power level in areas where the general public would be exposed is far below these levels and therefore should meet the test of a "reasonable" level of radiation.

12.2 SPECTRUM

12.2.1 Frequency Assignments to Fixed Satellite Services

Figure 12.1 and Table 12.1 show the principal frequency assignments for microterminals.The fixed satellite service (FSS) is an ITU definition that was adopted some years ago when earth stations could be classified into either the fixed (permanently mounted to the ground) or mobile categories. In recent years, an unofficial category (transportable) has evolved. These use the FSS frequencies and are licensed as though they were a FSS earth station, but they have no fixed or permanent address and are given special developmental licenses. The number of these transportable stations has now passed the 100 mark and is climbing rapidly. The principal users of these transportable stations are television stations and networks that wish to gather sporting events, news, and so on, from distances beyond the line of sight which is presently covered by their electronic news gathering facilities. Trucks are equipped with foldaway antennas (see Figure 12.2).

12.2.2 Additional Assignments

There is a growing concern that the existing frequency assignments for microterminals will be filled in the foreseeable future. This is because most of the systems are rather spectrum inefficient. The number of spread-spectrum networks per 36-MHz transponder is measured in terms of a few tens. This is substantially different from the high density of the trunk FDMA and TDMA and the companded single-sideband amplitude-modulation systems. These systems can accommodate from 900 to 6000 one-way voice channels per transponder simultaneously.

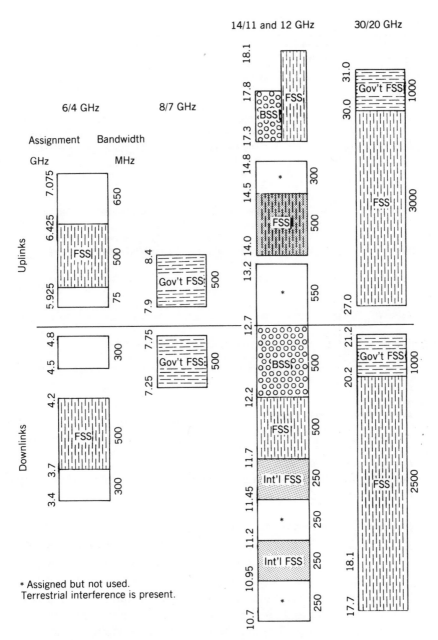

Figure 12.1. Western hemisphere frequency assignments.

TABLE 12.1 Comparison of the Bands

Band(GHz)	Bandwidth (MHz)	Uses	Rain[a]	Sources of Interferences
1.6[b]/1.5[c]	17 each	Mobile satellite	n/a	None
2.5[bc]	80	U.S.: none	n/a	IFTS and MDS
		Other: Broadcasting satellite		
4[c]	500[d]	Fixed satellite	n/a	Terrestrial
6[b]	500[d]	Fixed satellite	n/a	Terrestrial
7[c]	500	Government satellite	Low	
8[b]	500	Government satellite	Low	
11[c]	500	International satellite	Moderate	Terrestrial
12[c]	500	Fixed satellite	Moderate	None
DBS[c]	500	Broadcasting satellite	Moderate	None
14[b]	500	Domestic and international satellite	Moderate	None
17[b]	500	Broadcasting satellite	Moderate	Limited terrestrial
20[c]	2500	Fixed satellite	Higher	
30[b]	2500	Fixed satellite	Higher	

[a]n/a, not applicable. Rain loss.
[b]Uplink (earth to space).
[c]Downlink (space to earth).
[d]Expansion possibilities (with limitations).

For this reason there is a growing awareness that the U.S. direct-broadcast satellite frequency band between 12.2 and 12.7 GHz (with a matching uplink at 17.3–17.8 GHz) offers a number of advantages:

1. Orbital separation of the satellites (9°)
2. Technology that permits high-powered (100–200 W) power amplifiers
3. Authorization to use these high power links for nonvideo services provided that this does not exceed 50% of the satellite's capacity
4. Growing perception in the United States that direct broadcasting is a very expensive and risky business (especially with competition from videocassette recorders and the video channels on the existing satellite services at lower frequencies)

There are some secondary terrestrial users of these frequencies (primarily in the public safety sector). If satellite services are established, these services will have to move. If no satellite services are initiated, these frequencies may revert to all terrestrial uses.

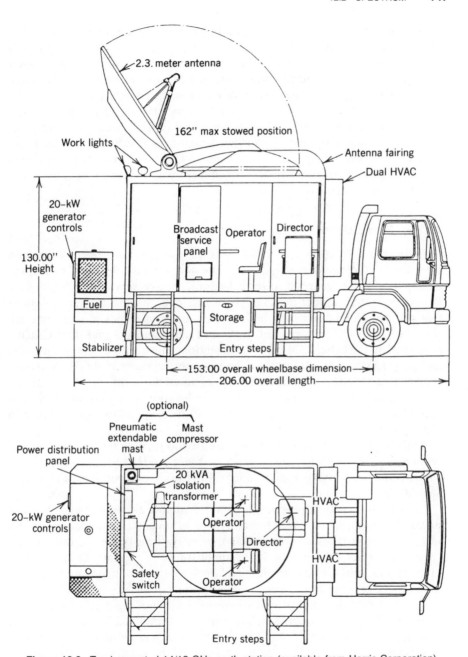

Figure 12.2. Truck-mounted 14/12-GHz earth station (available from Harris Corporation).

As is well recognized in Europe and Japan, the K_a-band frequencies around 20 GHz (with uplinks at 30 GHz) offer the potential for further expansion of these microterminals. At present, the technology is not sufficient, so a U.S.-government-funded program [Advanced Communications Technology Satellite (ACTS)] is seeking to provide experience. There are two drawbacks to the use of these frequencies: a less advanced state of technology and increased rainfall sensitivity. The major advantages are a wide presently uncongested frequency band. As further research is conducted and as these frequencies enter use (internationally) and in U.S. military applications (particularly at 44/20 GHz for Milstar), equipment will eventually become available for commercial use.

12.2.3 Mixed Uses

As indicated in Section 12.2.1, some of these bands have been authorized for both satellite and terrestrial uses. The 4- and 6-GHz satellite bands, for example, are also assigned to terrestrial microwave services. In general, it is necessary to conduct an analysis of the local microwave environment (usually, by a computer data base search, but sometimes in situ tests are needed). The FCC requires notification of existing users and their concurrence before licensing a 6/4-GHz earth station. This can take several months and from $200 to $20,000 per site. The two-way Equatorial 6/4-GHz terminals require frequency coordination. This has limited the ability for conventional small C-band terminals to go to center-city locations. Equatorial, with their spread-spectrum techniques, has successfully avoided the interference in most instances.

The receive-only Equatorial terminals are too small to be coordinated and registered. They must take their chances. Since 4-GHz terrestrial microwave systems may eventually vanish (due to fiber optic competition), the 4-GHz satellite band may increase in value.

Multiple uses may appear, even within a satellite exclusive band. It is not unusual to find a satellite that has television in several transponders, TDMA in two or more transponders, FDMA in many transponders, and spread-spectrum services along with SCPC applications (see Figure 12.3). The services require different bandwidths and power densities, which must accommodate one another in the same and adjacent satellites.

The Federal Communication Commission does not regulate what uses are made of a fixed satellite or which transponders are used for what services. It is up to the individual users to accommodate one another. In those few instances where this is impossible and one of the parties feels that one of the other users is violating an FCC rule, it may appeal to the FCC to require the other party to adjust their transmissions. Considering all the possible interference conditions, this is rarely necessary because the individual operators will usually work together to resolve the problem.

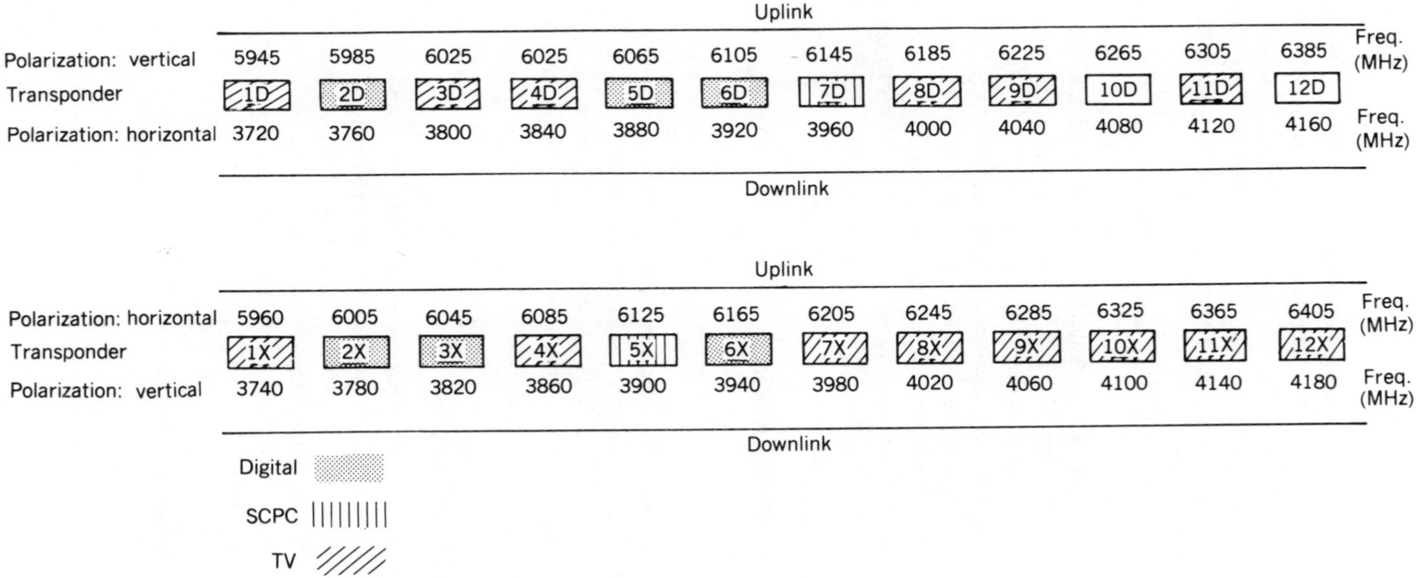

Uplink

Freq. (MHz)

Polarization: vertical	5945	5985	6025	6025	6065	6105	6145	6185	6225	6265	6305	6385
Transponder	1D	2D	3D	4D	5D	6D	7D	8D	9D	10D	11D	12D
Polarization: horizontal	3720	3760	3800	3840	3880	3920	3960	4000	4040	4080	4120	4160

Downlink

Freq. (MHz)

Uplink

Freq. (MHz)

Polarization: horizontal	5960	6005	6045	6085	6125	6165	6205	6245	6285	6325	6365	6405
Transponder	1X	2X	3X	4X	5X	6X	7X	8X	9X	10X	11X	12X
Polarization: vertical	3740	3780	3820	3860	3900	3940	3980	4020	4060	4100	4140	4180

Downlink

Freq. (MHz)

Digital
SCPC
TV

Figure 12.3. Mixed uses of *Westar V*.

149

12.3 POWER FLUX DENSITY

In frequency bands where satellites and terrestrial services are both present, there are limitations on how much power the satellite can place on the earth. The value varies with the elevation angle. The most stringent condition is at the 0° elevation angle, where the satellite transmits directly into a terrestrial microwave point-to-point beam. The power level is generally in terms of watts in the worst 4 kHz.

Table 12.2 shows the power flux density limitations for the various satellite bands. These are established by the International Telecommunications Union for Region 2 (North and South America). In the band 11.7–12.2 GHz there is the additional limitation of 53 dBW eirp per *video* channel. This was a level established to accommodate the Canadian Anik-C series of satellites that were providing video via spot beams in these frequency bands. As shown in Figure 12.4, one in-orbit SBS satellite has a power level close to this limit. If the service is not television, this limitation does not apply in the band from 11.7 to 12.2 GHz. Digital and voice services therefore escape this limitation.

12.4 ORBIT CONSIDERATIONS

12.4.1 Arcs

Figure 12.5 shows the orbital arcs occupied by the various North American satellites. There is no boundary between the North Atlantic and domestic arcs. The locations of the U.S., Canadian, and Mexican satellites have been agreed upon in multilateral meetings. Figure 12.6 shows the coverage of the U.S. population.

12.4.2 Orbital Crowding

To accommodate as many requests as possible for domestic satellites, the Federal Communication Commission has ordered that U.S. fixed satellites eventually be placed on a 2° orbital separation. This separation amounts to approximately 750 km along the orbital path.

If an earth station were located at the center of the earth, the satellites

TABLE 12.2 Power Flux Density Limits

Band	GHz	Power Flux Density Limit at 0°
C	3.7– 4.2	-152 dBW/m^2 (in 4 kHz)
K$_u$	11.7–12.2	No limit (53 dBW/*TV* channel)
K$_a$	17.7–20.2	-115 dBW/m^2 (in 1MHz)

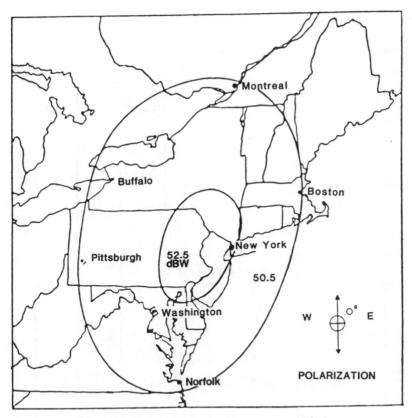

Figure 12.4. East spot beam EIRP contour (SBS-4).

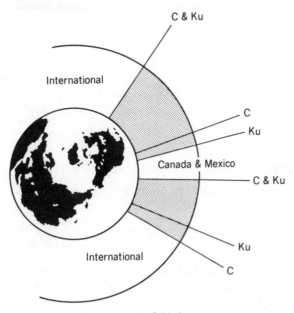

Figure 12.5. Orbital arcs.

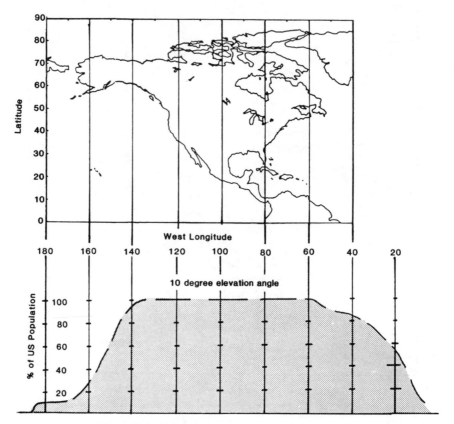

Figure 12.6. U.S. coverage from various orbit locations.

would appear to be 2.0° apart. Since it is located on the surface of the earth, the actual angle (as seen by the earth station) appears slightly larger (typically 2.2°). The precise angle depends on the location of the earth station. These angles are important due to the earth station antenna sidelobes (see Chapter 10).

12.4.3 Assignments

Table 12.3 shows the orbital locations as assigned by the Federal Communications Commission for domestic U.S. satellites.

12.4.4 Polarizations

Generally, the signal launched from the satellite is polarized to be either parallel or perpendicular to the equatorial plane. In two satellite systems (RCA's Satcom K and GTE's Gstar) transmission orientations are rotated

TABLE 12.3 Numeric Listing of U.S. Satellite Locations[a]

Future Location	FCC Name	Common Name	Company	GHz	8/87 Location
62	SBS 6	SBS-6	IBM	14/12	
62	Satcom 7	Satcom VII	GE	6/4	
64	ASC 3–4	ASC 3–4	Contel-ASC	6/4 and 14/12	
[b]	RCA C	Satcom K3	GE	14/12	
67	Unassigned			6/4	
69	Spacenet 2	Spacenet II	GTE	6/4 and 14/12	69
71	Galaxy A	Galaxy H5	GM/Hughes	14/12	
72	Satcom 2R	Satcom IIR	GE	6/4	72
73	Westar A	Westar A	WU	14/12	
74	Galaxy 2	Galaxy II	GM/Hughes	6/4	74
75	Unassigned			14/12	
101[c]	Comstar	Comstar D2/D4	Comsat	6/4	76
77	FedEx A	Expresstar I	Federal Express	14/12	
79	Unassigned			14/12	
79	Westar 3	Westar III	WU	6/4	91
81	RCA B	Satcom K2	GE	14/12	81
81	Satcom 4	Satcom IV	GE	6/4	83
83	ASC 2	ASC 2	Contel-ASC	6/4 and 14/12	
[b]	RCA A	Satcom K1	GE	14/12	85
85	Telstar 3	Telstar 302	AT&T Communications	6/4	86
87	Spacenet 3	Spacenet III	GTE	6/4 and 14/12	
89	Unassigned			6/4	
89	Unassigned			14/12	
91	SBS 4	SBS-4	IBM	14/12	91
91	Westar 6	Westar VI-S	WU	6/4	
93	Unassigned			6/4 and 14/12	
95	Galaxy 3	Galaxy III	GM/Hughes	6/4	93.5
95	SBS 3	SBS-3	MCI	14/12	95
75[c]	SBS 2	SBS-2	MCI	14/12	[b]
97	Telstar 2	Telstar 301	AT&T Communications	6/4	96
75[c]	SBS 1	SBS-1	Comsat	14/12	[b]
99	Westar 4	Westar IV	WU	6/4	99
101	Unassigned			6/4 and 14/12	
103	Gstar 1	Gstar I	GTE	14/12	103
105	Gstar 2	Gstar II	GTE	14/12	105
120	Spacenet 1	Spacenet II	GTE	6/4 and 14/12	120
122	SBS 5	SBS-5	IBM	14/12	
122	Unassigned			6/4	
124	FedEx B	Expresstar I	Federal Express	14/12	
124	Westar 5	Westar V	WU	6/4	122.5
126	Unassigned			14/12	
126	Telstar 1	Telstar 303	AT&T Communications	6/4	125
128	ASC 1	ASC-1	Contel-ASC	6/4 and 14/12	128
130	Galaxy B	Galaxy H6	GM/Hughes	14/12	
130	Satcom 3R	Satcom IIIR	GE	6/4	131
132	Galaxy 1	Galaxy I	GM/Hughes	6/4	134
132	Westar B	Westar B	WU	14/12	

Continued

TABLE 12.3 Continued

Future Location	FCC Name	Common Name	Company	GHz	8/87 Location
134	Unassigned			14/12	
134	Unassigned			6/4	
136	*Spacenet 4*	*Spacenet IV*	GTE	6/4 and 14/12	
136	*Gstar 3*	*Gstar III*	GTE	14/12	
138	*Satcom 1R*	*Satcom IR*	GE	6/4	139
140	*Galaxy 4*	*Galaxy IV*	GM/Hughes	6/4	
142	*Aurora 1*	*(Satcom V)*	Alascom	6/4	143
144	Unassigned			6/4	
146	Unassigned			6/4	

[a]Subject to change. As of August 1987.
[b]*Satcom K1* will move to 67 W after *K3* is launched. *K4* may join *K3* at 85 W.
[c]Per Comsat request.

slightly. When dealing with adjacent satellite interference situations, these rotations increase the amount of interference. The worst situation, however, is when two adjacent satellites have similar frequency and polarization assignments. Figure 12.7 shows a portion of the orbital arc and the polarization assignments at the frequency of the uplink. Figure 12.8 shows the situation for the downlink. Dual polarization satellites are designated by the cross symbol.

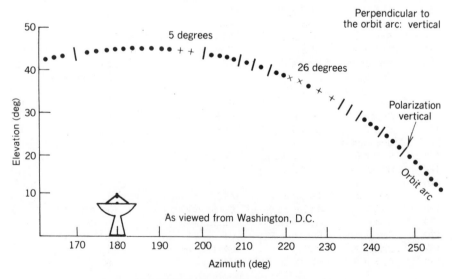

Figure 12.7. Uplink polarization plans.

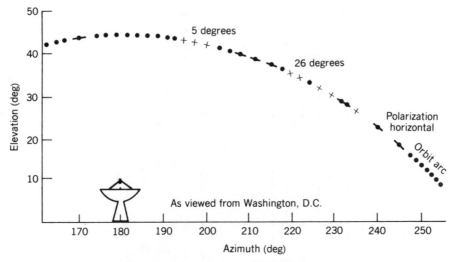

Figure 12.8. Downlink polarization plans.

12.4.5   Channel Assignments

Most satellites at C-band have a common 36-MHz frequency assignment method. The polarization plans may be reversed from one another, but generally their center transponder frequencies are identical. This situation got more complex when 72-MHz transponders were introduced, but even these tended to perpetuate the original assignments as far as possible. See Figure 12.9 for an example.

In the case of K_u-band, there is no universal transponder assignment method in the United States, Canada, or Mexico. Several methods are similar but have subtle differences. Variations are present in the bandwidth (36, 43, 54, 72, and 108 MHz are already in use). Even between systems using the same transponder bandwidth, the center frequencies differ.

12.4.6   Power Levels

Table 12.4 also shows the power amplifier ratings and typical equivalent isotropically radiated power levels for the K_u-band satellites.

12.4.7   Orbital Spacing

Table 12.5 shows how the spacing between satellites has changed over a period of time. The FCC eventually plans to have all U.S. domestic satellites spaced at 2° (for U.S. users) in both the C- and K_u-bands. The direct broadcast band (12.2–12.7 GHz) is planned for a 9° separation for satellites serving the

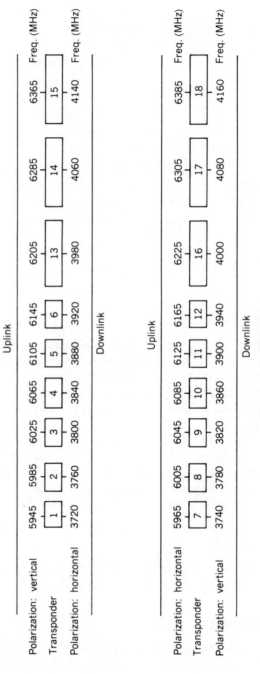

Figure 12.9. ASC 1 frequency plan.

TABLE 12.4 North American Domestic Satellites

Name	Operator	Number of Transponders per Satellite	Total Bandwidth (MHz)	Uplink (GHz)	G/Ts (dBi/K) at Edge	Downlink (GHz)	EIRP (dBW) at Edge
ASC 1 and ASC 2	American Satellite Co.	18	1296	5.925–6.425	−4	3.7–4.2	34 and 36
		6		14.0–14.5	−2	11.7–12.2	42
Anik-C1 to C3	Telesat Canada	16	864	14.0–14.5	+3	11.7–12.2	31–47
Anik-D1 to D2	Telesat Canada	24	864	5.925–6.425	−6	3.7–4.2	36
Aurora I (formerly Satcom V)	Alascom, Inc.	24	864	5.925–6.425	−6	3.7–4.2	34
Aurora IR	Alascom, Inc.	24	864	5.925–6.425	−4	3.7–4.2	34–38
Galaxy I to III	Hughes Communications Galaxy, Inc.	24	864	5.925–6.425	−3.4	3.7–4.2	34.5
Gstar I to III	GTE Spacenet	16	864	14.0–14.5	+1.6	11.7–12.2	43.1
Morelos 1 and 2	Mexico	18	1008	5.925–6.425	−9	3.7–4.2	35 and 44
		4		14.0–14.5		11.7–12.2	
Satcom IIIR and IV	GE Americom	24	864	5.925–6.425	−6	3.7–4.2	32 and 34
Satcom IR and IIR	GE Americom	24	864	5.925–6.425	−6	3.7–4.2	34
Satcom K1 and K2	GE Americom	16	864	14.0–14.5	0	11.7–12.2	43–47
SBS 1 to 3	Comsat	10	430	14.0–14.5	+1.8	11.7–12.2	37–43.8
SBS 4	IBM	10	430	14.0–14.5	+1.8	11.7–12.2	37–50
SBS 5 and 6	IBM	14	870	14.0–14.5	−1	11.7–12.2	47–49
Spacenet 1 to 3R	GTE Spacenet	12	1296	5.925–6.425	−5	3.7–4.2	34 and 36
		6		14.0–14.5	−5	11.7–12.2	39
Telstar 301 to 303	AT&T Communications	24	864	5.925–6.425	−5	3.7–4.2	32–36
Westar III	Western Union	12	432	5.925–6.425	−7.4	3.7–4.2	33
Westar IV and V	Western Union	24	864	5.925–6.425	−1	3.7–4.2	34
Westar VI-S	Western Union	24	864	5.925–6.425	−1	3.7–4.2	34

TABLE 12.5   Orbital Spacing

Year	Separation to Nearest Cofrequency Commercial Satellite (deg)[a]	Band	Notes
1963	360	C	*Syncom 1*
1965	360	C	*Early Bird (Intelsat I)*
1967	180 (E)	C	*Intelsat II–F2*
	20 (E)	C	*Intelsat II–F3*
	10 (E)	C	*Intelsat II–F4*
1970	8 (E)	C	*Intelsat IV–F2*
1980	3	C	*Intelsat IVA–F3*
1981	3	K_u	*SBS-2*
1982	2	K_u	*SBS-3*
1983	2	C	*Galaxy II*

[a](E), Estimated separation and date.

same areas. Additional satellites are interspersed with beams covering other parts of the earth.

Since the beamwidth (interference rejection) of an earth station antenna is inversely proportional to the antenna diameter, going from a 2° orbital separation to a 9° orbital separation could permit antennas to be two-ninths of the size of present antennas and still have the same adjacent satellite interference level. This would typically reduce a 1-m antenna to 22 cm (about 9 in.). An antenna this size is unobtrusive and may even be placed on a desk or table near a window. The window must not be coated with a solar reflector substance (e.g., a mirror film).

Practical limitations restrict the minimum antenna size. These involve the power levels in the satellite and the microterminal. Both power amplifiers would need to be increased substantially [by $(9/2)^2$] if the smallest microterminal antennas were used.

12.5   INTERFERENCE

12.5.1   Registration of Earth Stations

Historically, the Federal Communications Commission has required the registration and licensing of earth stations that transmit. Receive-only earth stations do not require licensing, but if the user of the station wanted to protect against future terrestrial interference (from microwave links), registration was advised. The FCC had a minimum receive-only antenna diameter below which they would not accept registration and licensing because the antenna beam was so broad that it imposed undue limitations on terrestrial microwave services.

In 1986, the FCC was requested to eliminate registration and licensing of two-way business earth stations provided that a type acceptance of each earth station class and model would be required in lieu of the licensing. The

FCC agreed to this suggestion. The type acceptances are limited to the satellite exclusive bands (11.7–12.2 and 14.0–14.5 GHz).

The FCC permits unlicensed receive-only Intelsat Intelnet I stations. Stations transmitting at 6 GHz still need to be registered. Temporary authorizations permit these stations to operate during coordination provided that they do not produce interference into other terrestrial services.

12.5.2 Terrestrial Interference

Figure 12.10 shows the type of interference that can be experienced by an earth station. It also shows that it is possible to provide interference shielding

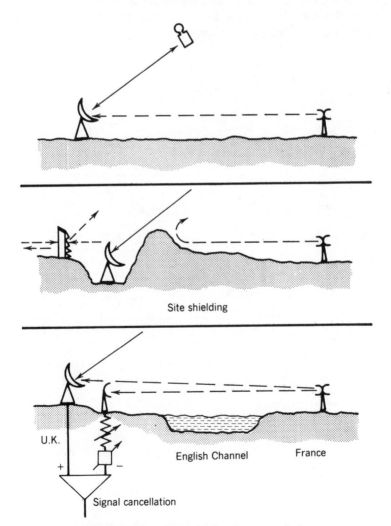

Figure 12.10. Earth station interference reduction.

to minimize these effects. These techniques are used at C-band (6/4 GHz) and are not needed at domestic K_u-band frequencies.

12.6 RADIATION CONCERNS

The Occupational Safety and Health Administration's limitation on personnel exposure to electromagnetic radiation is 10 mW/cm^2 per 6 hours.

In all but the most powerful earth stations, this flux level is exceeded only within the cone of the main beam area. Since this beam is elevated above the earth (typically, 10° or more), it generally is not a concern because personnel will be out of the beam. Chain link fences are often used around the transmitting antenna to prevent accidental exposure to people or animals walking by the antenna. A more important reason for the fence is security (to keep the curious and intruders out).

TECHNICAL CONSIDERATIONS

This section deals with the common technical elements of all microterminal systems.

13.1 TRANSMISSION LINKS

Four links are involved in a typical two-way microterminal situation:

Inbound to hub
   Microterminal to satellite transponder
   Satellite transponder to hub
Outbound from hub
   Hub to satellite transponder
   Satellite transponder to microterminal

Each link has its own constraints. In general, the limitations are shown in Table 13.1.

13.2 INBOUND TO HUB

13.2.1 Microterminal to Satellite Transponder

The microterminals do not have enough radiated power collectively to saturate the transponder. This link is bandwidth limited. Table 14.1 and Figure 14.2 show the number of channels (frequency slots) that can be derived as

TABLE 13.1 Link Constraints

Link	Constraints		Transponder Notes
	Earth Station	Satellite	
Inbound to hub			
Microterminal to satellite	Power and antenna area	Bandwidth	Multicarrier transponder
Satellite to hub	Interference	Bandwidth	Multicarrier transponder
Outbound from hub			
Hub to satellite	—	Power	Single or multicarrier
Satellite to microterminal	Noise and antenna area	Power	Single or multicarrier

a function of the bandwidths of the microterminal transmission and the transponder for FDMA service.

Using binary phase shift keying (BPSK), at least one bit per hertz (of bandwidth) may be achieved; 56 kb/s requires approximately 45–60 Hz. Quadrature phase shift keying (QPSK) can achieve approximately 2 bits per hertz. When space capacity becomes limited, we expect networks to convert from BPSK to QPSK. Still higher-order (M-ary phase shift keying) systems are available for further reductions in the bandwidth, but the equipment is costly and less reliable. Table 13.2 shows a typical 56-kb/s link into the hub. Other data rates (see Figure 2.10) are also used. These require proportionately greater bandwidths.

13.2.2 Satellite Transponder to Hub

The hub station has the advantage of a large antenna. The hub station is noise limited (by the low-noise receiver and the sky-noise temperatures). This is overcome by the use of a large earth station antenna to achieve the required figure merit (G/T). The primary limitations on the hub station are rain attenuation and accidental interference in the inbound uplink that would swamp the many low-level inbound carriers. Table 13.3 shows the typical downlink-to-hub link budget.

13.3 OUTBOUND FROM THE HUB

13.3.1 Hub to Satellite Transponder

The hub uplink is shown in Table 13.4. *Gstar II* has been chosen as a typical satellite in the class that includes *ASC 1*, *SBS-1* through *SBS-4*, and *Spacenet*. The uplink from the hub operates in the rate range for 0.5–21 Mb/s. The table is for the 512 kb/s, which is typical of the slower-speed services.

TABLE 13.2 Microterminal Uplink

Satellite name: Gstar II
Beam name: CONUS
Station location: Rain region D2
Service: 64 kb/s

	Weather		
	Clear	Rain	Unit
Frequency	14.25	14.25	GHz
High-power amplifier	1.00	1.00	W
Antenna diameter	1.80	1.80	m
Saturation EIRP	45.49	45.49	dBW
Off-axis loss	0.10	0.10	dB
Output backoff	0.00	0.00	dB
Station EIRP	45.39	45.39	dBW
Elevation angle	37.00	37.00	deg
Distance	38,400	38,400	km (approx.)
Path loss	207.21	207.21	dB
Rain loss	0.00	5.00	dB
Satellite illumination level	−117.31	−122.31	dBW/m^2
Satellite G/T on axis	3.89	3.89	dBi/K
Satellite margin	0.70	0.70	dB
Off-axis loss	2.00	2.00	dB
Carrier-to-thermal noise	−161	−161	dBW/K (C/T)
Boltzmann constant	228.60	228.60	dBW/K·Hz
C/kT	68	63	dBHz (C/N$_0$)
Noise bandwidth	0.045	0.045	MHz
C/N (up)	21	16	dB (C/kTB)

The high-power amplifier is 6.5 W per carrier. If a 10-Mb/s carrier were used, the HPA would need to be at least 250 W. At 20 Mb/s this climbs to 500W.

Solid-state power amplifiers are now available at 14 GHz at power levels up to 10 W. The 250-W transmitter (which may actually be 400 W at beginning of life) is most likely a traveling-wave tube.

The uplink carrier-to-noise ratio, under clear-sky conditions, is 34 dB. A 5-dB rain loss lowers this to 29 dB. Considering that the satellite transponder has a bandwidth between 27 and 72 MHz, these carriers occupy only a fraction of the total capacity available (about 0.4 and 8 MHz per carrier for the two cases). The transponder generally runs out of power long before the full bandwidth is consumed.

13.3.2 Satellite Transponder to Microterminal

This is the link in which power is the limitation. This is the critical path in a network. The most efficient operation is on a single-carrier basis. A 21-Mb/s TDM bit stream may need 10–20 MHz (depending on the modulation

TABLE 13.3 Downlink to Hub

Satellite name:	Gstar II			
Beam name:	CONUS			
Station location:	Rain region D2			
Service:	64 kb/s			

	Weather				
	Clear	Clear	Rain	Rain	Unit
Number of TWTAs	1	1	1	1	tube
Power level	20.0	20.0	20.0	20.0	W
Frequency	11.95	11.95	11.95	11.95	GHz
Saturation EIRP	44.5	44.5	44.5	44.5	dBW
Off-axis loss	4.0	4.0	4.0	4.0	dB
Output backoff	5.0	5.0	5.0	5.0	dB
Channels per transponder	600	1200	600	1200	at 64 kb/s
Power split	27.8	30.8	27.8	30.8	dB
Saturation EIRP per carrier	7.7	4.7	7.7	4.7	dBW
Elevation angle	37	37	37	37	deg
Distance	37,854	37,854	37,854	37,854	km (approx.)
Path loss at 12 GHz	205.6	205.6	205.6	205.6	dB
Rain loss	0.0	0.0	3.0	3.0	dB
Earth illumination level	-155.3	-158.3	-158.3	-161.3	dBW/m^2/45 kHz
Power flux density	-165.8	-168.8	—	—	dBW/m^2/45 kHz
Earth station antenna diameter	9.0	9.0	9.0	9.0	m
Antenna efficiency	60	60	60	60	%
Antenna gain on axis	58.8	58.8	58.8	58.8	dBi
LNA temperature	225	225	225	225	K
Sky noise	30	30	150	150	K
System noise temperature	255	255	375	375	K
Earth station G/T	34.7	34.7	33.1	33.1	dBi/K
Earth station maintenance margin	0.7	0.7	0.7	0.7	dB
Carrier-to-thermal noise	-164.3	-167.3	-169.0	-172.0	dBW/K (C/T)
Boltzmann constant	228.6	228.6	228.6	228.6	dBW/K·Hz
C/KT	64.3	61.3	59.6	56.6	dBHz (C/N$_0$)
Noise bandwidth	0.045	0.045	0.045	0.045	MHz
C/N (down)	17.8	14.8	13.1	10.1	dB (C/KTB)
C/N (up)	21.0	21.0	21.0	21.0	dB Clear Uplink
C/IM (sat.)	32.0	32.0	32.0	32.0	dB
C/I (down)	35.0	35.0	35.0	35.0	dB
C/(N + I) total	15.9	13.7	12.4	9.7	dB
10^{-6} BER BPSK	10.50	10.50	10.50	10.50	dB
Margin	5.43	3.25	1.89	-0.78	dB

TABLE 13.4   Hub Uplink

Satellite name: Gstar II
Beam name: CONUS
Station location: Rain region D2
Service: 512 kb/s

	Weather		
	Clear	Rain	Unit
Frequency	14.25	14.25	GHz
High-power amplifier	6.50	6.50	W
Antenna diameter	9.00	9.00	m
Saturation EIRP	67.59	67.59	dBW
Off-axis loss	0.10	0.10	dB
Output backoff	0.00	0.00	dB
Station EIRP	67.49	67.49	dBW
Elevation angle	37.00	37.00	deg
Distance	38,400	38,400	km (approx.)
Path loss	207.21	207.21	dB
Rain loss	0.00	5.00	dB
Saturation illumination level	-95.20	-100.20	dBW/m^2
Satellite G/T on axis	3.89	3.89	dBi/K
Satellite margin	0.70	0.70	dB
Off-axis loss	2.00	2.00	dB
Carrier-to-thermal noise	-139	-144	dBW/K (C/T)
Boltzmann constant	228.60	228.60	dBW/K·Hz
C/kT	90	85	dBHz (C/N$_0$)
Noise bandwidth	0.400	0.400	MHz
C/N (up)[a]	34	29	dB (C/kTB)

[a]10-Mb/s signal will require a 250-W transmitter to achieve the same C/N (up) performance.

method). This suggests that many narrow-band transponders (e.g., 27 MHz) may be more desirable than a few broader-band transponders per satellite.

Table 13.5 shows the downlink to the microterminal. Four cases are shown, two for the 512-kb/s and two for the 10-Mb/s cases. The first two columns represent the clear-sky condition in the downlink, and the last two accommodate all but 0.5% of the situations in rain region D2 (see Figure 10.11).

The on-axis saturated EIRP is 44.5 dBW (as shown in Figure 13.1). If coverage is assumed to be the 42.5–44.5 dBW range (the dotted area of Figure 13.1), the performance will be as shown in Table 13.5 for 1.8-m antennas. The table is also valid within the line-shaded area provided that the microterminal antenna diameter is at least 2.5 m.

If multiple carriers are present in the transponder's TWTA, its output power must be reduced by approximately 4 dB. If only a single carrier is present, no output backoff is necessary. In the single-carrier case it may be advisable to operate the transponder deliberately slightly beyond saturation to improve its ability to withstand fading in the uplink. Figure 13.2 shows

TABLE 13.5 Downlink to microterminal

Satellite name: Gstar II (Conventional TWTA)
Beam name: CONUS
Station location: Rain region D2
Service: See below

	Weather				
	Clear	Clear	Rain	Rain	Unit
Bit rate	512 kb/s	10 Mb/s	512 kb/s	10 Mb/s	
Voice channels per					
transponder	64	156	64	156	at 64 kb/s
Power level	20.0	20.0	20.0	20.0	W
Frequency	11.95	11.95	11.95	11.95	GHz
Saturation EIRP	44.5	44.5	44.5	44.5	dBW
Off-axis loss	2.0	2.0	2.0	2.0	dB
Output backoff	4.0	0.0	4.0	0.0	dB
Carriers per transponder	8	1	8	1	at bit rate
Power split per carrier	9.0	0.0	9.0	0.0	dB
Saturation EIRP per carrier	29.5	42.5	29.5	42.5	dBW
Elevation angle	37	37	37	37	deg
Distance	37,854	37,854	37,854	37,854	km (approx.)
Path loss at 12 GHz	205.6	205.6	205.6	205.6	dB
Rain loss	0.0	0.0	3.0	3.0	dB
Earth illumination level	−133.5	−120.5	−136.5	−123.5	dBW/m^2
Earth station antenna					
diameter	1.8	1.8	1.8	1.8	m
Antenna efficiency	60	60	60	60	%
Antenna gain on axis	44.8	44.8	44.8	44.8	dBi
LNA temperature	225	225	225	225	K
Sky noise	30	30	150	150	K
System noise temperature	255	255	375	375	K
Earth station G/T	20.8	20.8	19.1	19.1	dBi/K
Earth station maintenance					
margin	0.7	0.7	0.7	0.7	dB
Carrier-to-thermal noise	−156.5	−143.5	−161.2	−148.2	dBW/K (C/T)
Boltzmann constant	228.6	228.6	228.6	228.6	dBW/K·Hz
C/KT	72.1	85.1	67.4	80.4	dBHz (C/N$_0$)
Noise bandwidth	0.400	8.000	0.400	8.000	MHz
C/N (down)	16.1	16.1	11.4	11.4	dB (C/KTB)
C/N (up)	34.0	34.0	34.0	34.0	dB Clear Uplink
C/IM (sat.)	32.0	32.0	32.0	32.0	dB
C/I (down)	35.0	35.0	35.0	35.0	dB
C/(N + I) total	15.8	15.9	11.3	11.3	dB
10^{-6} BER BPSK	10.50	10.50	10.50	10.50	dB
Margin	5.34	5.36	0.82	0.84	dB

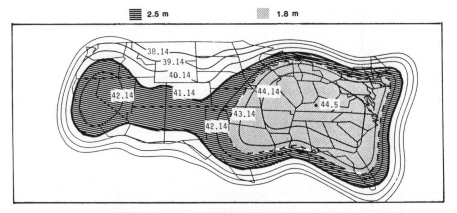

Figure 13.1. Typical K$_u$-band CONUS coverage.

the advantage of overdriving the traveling-wave tube. A 5.5-dB change in the uplink power results in a 0.5-dB change in the downlink EIRP.

When multiple carriers pass through a traveling-wave tube amplifier, the relative power output of each carrier must be reduced by at least 4 dB. Figure 13.3 shows how the third-order intermodulation (IM) products increase as the nonlinear tube is driven harder. At saturation the carrier-to-IM interference ratio is only slightly better than 6 dB, in the case shown. Linearizers

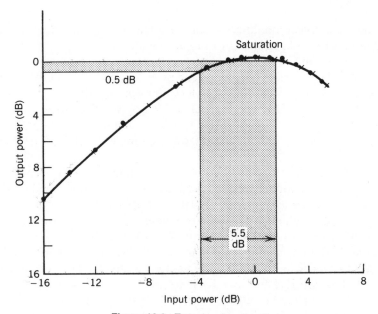

Figure 13.2. Transponder operation.

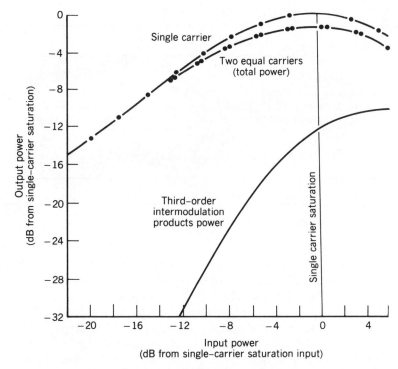

Figure 13.3. TWTA intermodulation.

and solid-state power amplifiers reduce this effect, but some output backoff is still necessary to keep the C/I ratio acceptable.

Returning to Table 13.5, the number of channels per transponder is shown. At 512 kb/s, eight carriers (channels) can be accommodated before the transponder runs out of power. Eight carriers represent a 9-dB power split. This reduces the satellite EIRP/per carrier to 29.5 dBW. The single carrier per transponder (at 10 Mb/s in an 8-MHz bandwidth) requires no output backoff or power split and therefore can utilize the entire 42.5 dBW off-axis power capability.

The downlink rain loss for a 99.5% case is approximately 3 dB in rain region D2. For this model we used a 1.8-m antenna diameter with a gain of 44.8 dBi. The low-noise amplifier (LNB or LNC) was set at 225 K. The sky-noise temperature changes between the clear-sky and rain condition. This directly effects the earth station figure of merit (G/T).

The earth station maintenance margin includes mispointing, foundation settling, receiver aging, wind, and temperature effects. The resulting C/kT (or C/N_0) ratios are 72.1 and 85.1 dBHz for the two bit rates in clear weather. In the rain these values become 67.4 and 80.4 dBHz, respectively. The carrier-to-noise ratio of the downlink is determined using the IF bandwidth (B).

TABLE 13.6   Receiver Noise Temperatures at K_u-Band

Element	Clear Sky		Heavy Rain	
Low-noise device[a] (K)	100	225	100	225
Antenna noise (K)	30	30	150	150
Total system noise (K)	130	255	250	375
dBK	21.1	24.1	24.0	25.7
1.2 m antenna gain (12 GHz) (dBi)	41	41	41	41
G/T (dBi/K)	19.9	16.9	17.0	15.3
Difference (dB)		3.0		1.7

[a]LNA, LNB, or LNC.

Since the downlink is the weakest link, the C/N (up), C/IM (sat), and C/I (down) result in only a modest change, which is shown as C/(N + I) total. A C/(N + I) of 10.5 dB is necessary for a bit error rate of 1 per million (1×10^{-6}). The final row shows the margin in the clear-sky and rain conditions for the two bit rates. The number of channels per transponder was adjusted to achieve a just operable condition during the heavy rain storm. As can be seen from Table 13.5 and Figure 14.3, this link is severely power limited.

At C-band this situation would be remedied (at least in part) by using a lower-noise-temperature (LNA). At K_u-band, the most critical link is between the hub and the receiving microterminal. When the microterminal uses a 1.2-m antenna the gain is low (3 dB below the 1.8-m antenna). The novice engineer may be tempted to reduce the noise temperature of the LNA/LNB or LNC to restore some of the G/T (figure of merit) sacrificed when the small antenna was selected.

It works fine at K_u-band under clear-sky conditions when the antenna (sky) temperature may be 30 K. The critical test of a K_u-band station awaits a rainstorm when the rain attenuation reduces the carrier (C-band). At the same time the antenna "sees" the raindrops which are at a near-ambient temperature. Under heavy rain the antenna (sky) noise temperature may rise into the range 100–200 K. Table 13.6 compares the system performance of a 100- and a 225-K LNA, LNB, or LNC. The difference between the two configurations *in rain* is only 1.7 dB.

13.4   SATELLITE TRANSPONDER CHOICES AVAILABLE

Table 13.7 compares the various K_u-band satellites used in the United States. For consistency we will continue to use the GSTAR because this satellite has a variety of configurations that can be compared on a consistant basis.

The individual transponders can be operated in either a linear or a limiter

TABLE 13.7   Comparison of K$_u$-Band Satellites

Name	Operator	K$_u$-Band Transponders Active	K$_u$-Band Transponders Spare	Bandwidth (MHz)	Transponder RF Watts[b]	Total RF Watts[a]
Anik-C	Telesat Canada	16	4	54	15	240
Anik-E	Telesat Canada	16[c]	—	54	—	—
ASC 1 and 2	American Satellite[a]	6[b]	1	72	16.2	296[a]
Spacenet I to III	GTE[a]	6[b]	1	72	16.5	297[a]
Gstar I to III	GTE	16	6	54	20 and 27	334
Morelos	Mexico	4[c]	2	108	19.4	225[a]
Satcom K1 and K2	GE Americom	16	6	54	40	640
Satcom K3	GE Americom	16	6	54	60	960
SBS-1 to 4	MCI and IBM	10	6	43	20	200
SBS-5	IBM	14	7	43 and 110[d]	20	280
SBS-6	IBM	19	12	43	41	780

[a]Sum of the rated powers of all active communications transmitters (including any at C-band).
[b]Also contains 18 C-band transponders.
[c]Also contains 24 C-band transponders.
[d]10 at 43 MHz + 4 at 110 MHz.

mode. The linear mode is desired for multicarrier (SCPC) operation. The limiter mode is designed for single-carrier-per-transponder operation where uplink rain attenuation can be mitigated by operating the carrier well into the limiter region.

A block diagram of the GSTAR communications subsystem is shown in Figure 13.4. A simplified block diagram showing the major elements of a generic transponder is shown in Figure 13.5 for the vertical receive–horizontal transmit polarization. A similar block diagram would apply to the horizontal receive–vertical transmit polarization.

The following is a description of the signal path through a transponder using Figure 13.5 as a reference.

A. All vertical polarized signals are picked up by the antenna and sent (1) to the satellite receiver (2).

B. The receiver frequency translates and amplifies all the signals and sends them to the input multiplexers (3). Only one of the multiplexers is shown.

C. The input multiplexers separate the signals by frequency into individual transponder channels. Assume, for example, that the signal we are tracing through the system has been filtered out by input multiplexers in group 1 for transmission through the second power amplifier from the left.

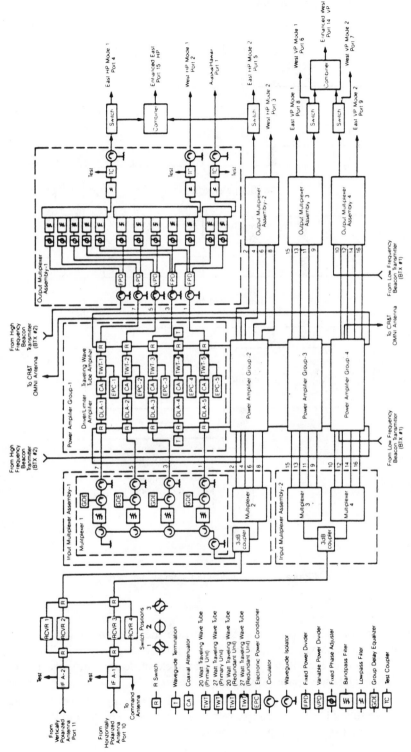

Figure 13.4. K$_u$-band satellite block diagram. Courtesy of GTE Spacenet.

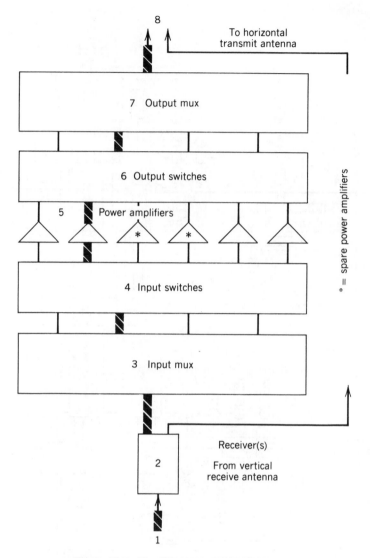

Figure 13.5. Simplified transponder diagram.

D. Input (4) and output (6) switches route the filtered signals from the input multiplexer (3) into the proper power amplifiers (5). There are spare power amplifiers provided (in the example shown, there are four active and two spare power amplifiers).

E. The power amplifier (5) amplifies the signal in a high-power traveling-wave tube (TWT) or solid-state power amplifier (SSPA) and then sends the signal to the output multiplexer (7).

F. The output multiplexer (7) combines all the signals from its power amplifiers. These are then combined with the other groups and sent to the horizontal transmit antenna (8).

13.4.1 Power Amplifiers

Most satellite transmitters (4 or 12 GHz) fall into three general classes. The traditional is the travelling-wave-tube amplifier (TWTA). Recently, these have been linearized as shown in Figure 13.6. The final type is a solid-state power amplifier.

As shown in Figure 13.6, an objective of a carrier-to-intermodulation distortion ratio of 20 dB can be achieved by an output backoff of the traveling-wave tube of 4–5 dB. The same objective may be achieved with a backoff of approximately 2 dB using a TWTA with a linearizer. This is also shown in Table 13.8, which is a recalculation of Table 13.5. The extra 2 dB from the linearizer is translated into additional channels. Solid-state power amplifiers are also linear devices and exhibit a similar increase in capacity.

13.4.2 Coverage

Figure 13.7 shows the coverage from a satellite into a spot beam in the western United States. Rain region C is included in this area. It is relatively dry compared to other parts of the nation. Table 13.9 shows that the capacity increases again. In this case it is slightly more advantageous to use the 512-kb/s channels instead of a 10-Mb/s system as shown in the second row ("channels per transponder"). Under these conditions, the power-limited downlink to the microterminal capacity (600 voice channels) is the same as one of the bandwidth-limited uplink cases (see Table 13.2). There is no absolute need to match the two link capacities, but it may lead to system econ-

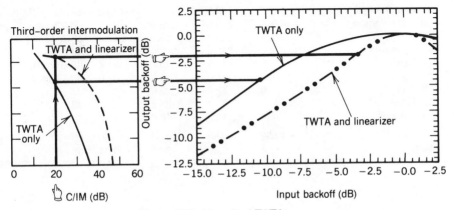

Figure 13.6. Linearized TWTAs.

TABLE 13.8   Downlink to Microterminal

Satellite name: Gstar II with TWTA and linearizer
Beam name: CONUS
Station location: Rain region D2
Rate: See below

	Weather				
	Clear	Clear	Rain	Rain	Unit
Bit rate	512 kb/s	10 Mb/s	512 kb/s	10 Mb/s	
Voice channels per transponder	120	156	120	156	at 64 kb/s
Power level	20.0	20.0	20.0	20.0	W
Frequency	11.95	11.95	11.95	11.95	GHz
Saturation EIRP	44.5	44.5	44.5	44.5	dBW
Off-axis loss	2.0	2.0	2.0	2.0	dB
Output backoff	2.0	0.0	2.0	0.0	dB
Carriers per transponder	15	1	15	1	at bit rate
Power split per carrier	11.8	0.0	11.8	0.0	dB
Saturation EIRP per carrier	28.7	42.5	28.7	42.5	dBW
Elevation angle	37	37	37	37	deg
Distance	37,854	37,854	37,854	37,854	km (approx.)
Path loss at 12 GHz	205.6	205.6	205.6	205.6	dB
Rain loss	0.0	0.0	3.0	3.0	dB
Earth illumination level	−134.3	−120.5	−137.3	−123.5	dBW/m^2/400 kHz
Earth station antenna diameter	1.8	1.8	1.8	1.8	m
Antenna efficiency	60	60	60	60	%
Antenna gain on axis	44.8	44.8	44.8	44.8	dBi
LNA temperature	225	225	225	225	K
Sky noise	30	30	150	150	K
System noise temperature	255	255	375	375	K
Earth station G/T	20.8	20.8	19.1	19.1	dBi/K
Earth station maintenance margin	0.7	0.7	0.7	0.7	dB
Carrier-to-thermal noise	−157.2	−143.5	−161.9	−148.2	dBW/K (C/T)
Boltzmann constant	228.6	228.6	228.6	228.6	dBW/K·Hz
C/KT	71.4	85.1	66.7	80.4	dBHz (C/N$_0$)
Noise bandwidth	0.400	8.000	0.400	8.000	MHz
C/N (down)	15.3	16.1	10.7	11.4	dB (C/KTB)
C/N (up)	34.0	34.0	34.0	34.0	dB Clear Uplink
C/IM (sat.)	32.0	32.0	32.0	32.0	dB
C/I (down)	35.0	35.0	35.0	35.0	dB
C/(N+I) total	15.1	15.9	10.6	11.3	dB
10^{-6} BER BPSK	10.50	10.50	10.50	10.50	dB
Margin	4.64	5.36	0.10	0.84	dB

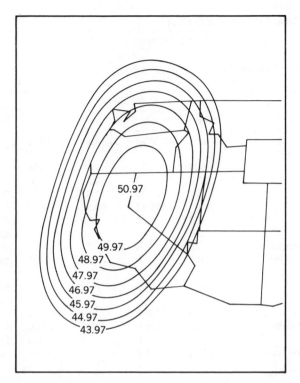

Figure 13.7. *Gstar II* west spot beam. Polarization is rotated 26° from horizontal and vertical. (From GTE Spacenet.)

omies if the two links use different transponders. A similar improvement may be obtained by using a higher-powered amplifier in the satellite (such as in the case of *Satcom K*).

13.5 EARTH STATION OPTIONS

13.5.1 Antenna Diameter

In the situation where the downlink to the microterminal is the limiting link (see Section 13.3.2), the amount of traffic through the final link is directly proportional to the area of the earth station antenna. This is illustrated in Table 13.10, which compares the channel capacity of a 1.8- and a 1.2-m earth station network. The ratio of the areas of the 1.8- and 1.2-m antennas are 2.25:1, which is almost the same as the ratio of the number of channels per transponder (112:48). The cost of the space segment for the downlink to the microterminal is therefore inversely proportional to the microterminal's antenna diameter.

TABLE 13.9   Downlink to Microterminal

Satellite name:	Gstar II with TWTA linearizer
Beam name:	West Spot
Station location:	Rain region C
Rate:	See below

	Weather				
	Clear	Clear	Rain	Rain	Unit
Bit rate	512 kb/s	10 Mb/s	512 kb/s	10 Mb/s	
Voice channels per transponder	600	156	600	156	at 64 kb/s
Power level	20.0	20.0	20.0	20.0	W
Frequency	11.95	11.95	11.95	11.95	GHz
Saturation EIRP	50.97	50.97	50.97	50.97	dBW
Off-axis loss	2.0	2.0	2.0	2.0	dB
Output backoff	2.0	0.0	2.0	0.0	dB
Carriers per transponder	75	5	75	5	at bit rate
Power split per carrier	18.8	7.0	18.8	7.0	dB
Saturation EIRP per carrier	28.2	42.0	28.2	42.0	dBW
Elevation angle	37	37	37	37	deg
Distance	37,854	37,854	37,854	37,854	km (approx.)
Path loss at 12 GHz	205.6	205.6	205.6	205.6	dB
Rain loss	0.0	0.0	2.0	2.0	dB
Earth illumination level	−134.8	−121.0	−136.8	−123.0	dBW/m^2/400 kHz
Earth station antenna diameter	1.8	1.8	1.8	1.8	m
Antenna efficiency	60	60	60	60	%
Antenna gain on axis	44.8	44.8	44.8	44.8	dBi
LNA temperature	225	225	225	225	K
Sky noise	30	30	100	100	K
System noise temperature	255	255	325	325	K
Earth station G/T	20.8	20.8	19.7	19.7	dBi/K
Earth station maintenance margin	0.7	0.7	0.7	0.7	dB
Carrier-to-thermal noise	−157.8	−144.0	−160.8	−147.1	dBW/K (C/T)
Boltzmann constant	228.6	228.6	228.6	228.6	dBW/K·Hz
C/KT	70.8	84.6	67.8	81.5	dBHz (C/N$_0$)
Noise bandwidth	0.400	8.000	0.400	8.000	MHz
C/N (down)	14.8	15.6	11.8	12.5	dB (C/KTB)
C/N (up)	34.0	34.0	34.0	34.0	dB Clear Up
C/IM (sat.)	32.0	32.0	32.0	32.0	dB
C/I (down)	35.0	35.0	35.0	35.0	dB
C/(N + I) total	14.6	15.4	11.7	12.4	dB
10^{-6} BER BPSK	10.50	10.50	10.50	10.50	dB
Margin	4.15	4.87	1.18	1.91	dB

TABLE 13.10 Downlink to Microterminal

Satellite name: Gstar II with TWTA linearizer
Beam name: CONUS
Station location: Rain region D2
Rate: 512 kb/s

	Weather				
	Clear	Clear	Rain	Rain	Unit
Bit rate	512 kb/s	512 Mb/s	512 kb/.	512 kb/s	
Voice channels per transponder	48	112	48	112	at 64 kb/s
Power level	20.0	20.0	20.0	20.0	W
Frequency	11.95	11.95	11.95	11.95	GHz
Saturation EIRP	44.50	44.50	44.50	44.50	dBW
Off-axis loss	2.0	2.0	2.0	2.0	dB
Output backoff	2.0	2.0	2.0	2.0	dB
Carriers per transponder	6	14	6	14	channels
Power split per carrier	7.8	11.5	7.8	11.5	dB
Saturation EIRP per carrier	32.7	29.0	32.7	29.0	dBW
Elevation angle	37	37	37	37	deg
Distance	37,854	37,854	37,854	37,854	km (approx.)
Path loss at 12 GHz	205.6	205.6	205.6	205.6	dB
Rain loss	0.0	0.0	3.0	3.0	dB
Earth illumination level	− 130.3	− 134.0	− 133.3	− 137.0	dBW/m^2/400 kHz
Earth station antenna diameter	1.2	1.8	1.2	1.8	m
Antenna efficiency	60	60	60	60	%
Antenna gain on axis	41.3	44.8	41.3	44.8	dBi
LNA temperature	225	225	225	225	K
Sky noise	30	30	150	150	K
System noise temperature	255	255	375	375	K
Earth station G/T	17.2	20.8	15.6	19.1	dBi/K
Earth station maintenance margin	0.7	0.7	0.7	0.7	dB
Carrier-to-thermal noise	− 156.8	− 156.9	− 161.5	− 161.6	dBW/K (C/T)
Boltzmann constant	228.6	228.6	228.6	228.6	dBW/K·Hz
C/KT	71.8	71.7	67.1	67.0	dBHz (C/N$_o$)
Noise bandwidth	0.400	0.400	0.400	0.400	MHz
C/N (down)	15.8	15.6	11.1	11.0	dB (C/KTB)
C/N (up)	34.0	34.0	34.0	34.0	dB Clear Up
C/IM (sat.)	32.0	32.0	32.0	32.0	dB
C/I (down)	35.0	35.0	35.0	35.0	dB
C/(N + I) total	15.6	15.4	11.0	10.9	dB
10^{-6} BER BPSK	10.50	10.50	10.50	10.50	dB
Margin	5.08	4.93	0.55	0.39	dB

These factors may force future earth stations to be closer to the 1.8-m size than the present 1.2-m antenna size. In areas where there is frequent heavy rain (such as in Florida or Italy), still larger antennas are necessary to provide an adequate margin in rain.

Antenna Pointing. With the larger antenna diameters, antenna pointing is a matter for concern. Figure 13.8 shows the "box" in which a geostationary K_u-band satellite is typically located. The dimensions about the normal (assigned) location are $\pm 0.05°$ north/south and east/west. This means that the box has a peak-to-peak $0.1° \times 0.1°$ dimension. The diagonal is $0.14°$.

Pointing a narrow-beam earth station antenna is not easy, even for a perfectly positioned satellite. An error of half of a degree becomes 187 km (302 statute miles) at the geostationary orbit.

Satellites are only briefly at their assigned location. During the rest of the time they are someplace within a tolerance box. If the satellite has an in-

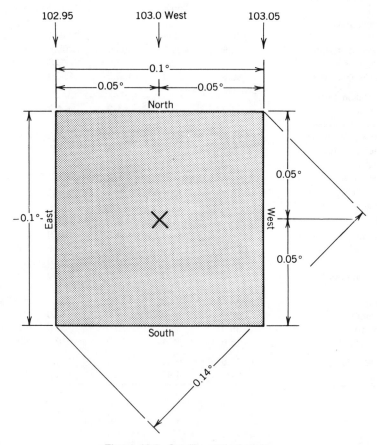

Figure 13.8. Satellite stationkeeping.

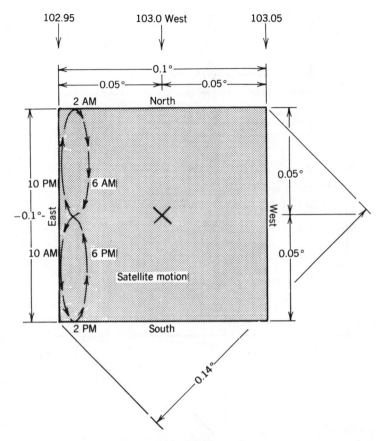

Figure 13.9. Satellite motion.

clination, the motion is described by a very thin figure eight (see Figure 13.9) which greatly exaggerates the width of the figure eight for the purpose of clarity. The period of the figure eight is one sidereal day (approximately 24 hours). We have arbitrarily labeled the times of the satellite location throughout this cyclical figure eight. The actual times for a particular satellite probably will be different (and will change due to stationkeeping maneuvers, drift, etc.).

Depending on when the antenna installer decides to make the final pointing to the satellite, there are a variety of locations to which the antenna might be pointed. If, for example, the pointing was done at 2 P.M., it would be pointed considerably to the south and east of the nominal location (the X in the center of the box).

Eventually, the satellite will be repositioned (by stationkeeping). Figure 13.10 shows one change of station location and a reduction in the inclination angle. The illustration is to show the change in satellite location and not

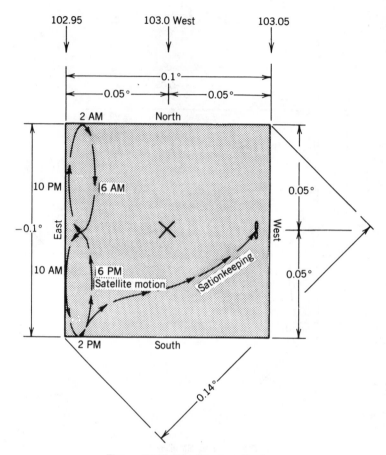

Figure 13.10. Satellite repositioning.

necessarily exactly how it goes from one position to the other. With the satellite at the west end of the box, there will be a loss in signal if the antenna is still pointing at the southeastern corner. It is thus possible to point the antenna precisely when the satellite is in the wrong place. As the satellite moves away from the point at which the antenna is pointed, the signal strength will decrease.

Figure 13.11 shows the half-power beamwidth versus antenna diameter. At best, only half of the indicated beamwidth applies in the case of the pointing errors. Figure 13.12 superimposes an initial pointing error of 0.2 beamwidth, and a ±0.05° satellite position accuracy, a worst case (wind/temperature and foundation shift of 0.3 beamwidth). This shows that for most K_u-band microterminal antennas these errors are not appreciable, but as the antenna approaches 5 m the errors consume more and more of the beamwidth

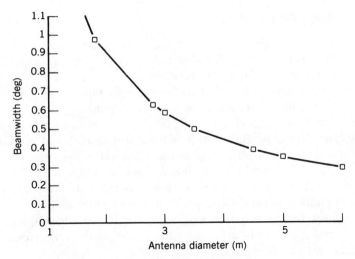

Figure 13.11. Antenna beamwidth (half-power beamwidth).

and therefore degrade the signal from the antenna. With still larger antennas, some form of tracking is needed. Tracking is expensive. Obviously, there is a trade-off between increased antenna diameter (and a better carrier-to-noise ratio), and a cost step due to tracking equipment to keep the antenna pointed at a moving satellite. At 12 GHz, the crossover is between 5 and 8 m.

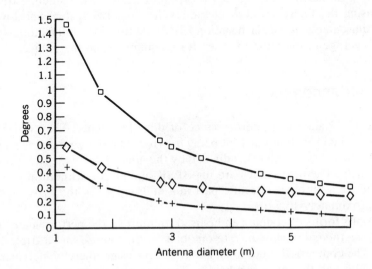

□, Beamwidth; +, initial and satellite motion; ◇, worst case

Figure 13.12. Beamwidth and errors.

13.5.2 Noise Temperature

As indicated above, the use of a low-noise temperature front end in a rainy climate is unwise. In a very dry (desertlike) area where rain is rare, a lower-noise-temperature amplifier may be considered. Since the market for these low-noise amplifiers (below 150 K) is presently limited, the price may be higher than the incremental cost of a larger antenna.

Generally speaking, K_u-band earth stations are noise and aperature limited. The designer has little control over noise and is constrained by economics, aesthetics, and ease of placement in the selection of the antenna diameter.

In the case of C-band microterminals, low-noise amplifiers in the range 70–80 K are readily available and may be advisable. In the case of spread-spectrum multiple access, the amount of noise contributed by the front end of the microterminal may be small compared to the adjacent satellite and intratransponder noise interference, and therefore the improvement in the total noise may be limited.

13.5.3 Power Amplifiers

Most microterminals require power amplifiers in the range 1–10 W. These are readily available in both vacuum-tube and solid-state power amplifiers. As the bit rate through the earth station is increased, the power amplifier rating must be increased in the same proportion. In the near future, this rate of increase will not be so great that the coming generations of solid-state power amplifiers cannot accommodate their needs. Eventually, it may be desirable to use spot or regional reception beams on the satellite (thereby increasing the figure of merit of the satellite receive system and lowering the requirements of the earth station EIRP). At this time, full national (CON-US) coverage of the nation is used for its simplicity in the uplink.

13.6 BIT ERROR RATES

Figure 13.13 shows the performance of digital systems. The vertical axis indicates the probability of 1 bit being in error. The horizontal axis plots the ratio of the energy per bit (E_b) divided by the noise density (N_0) as expressed in decibels. The heavy lines are for BPSK and QPSK without any coding. To achieve a bit error rate of 1 part in a million (10^{-6}), the E_b/N_0 needs to be approximately 10.5 dB.

Various forms of coding can be used to improve the performance. Three coding methods are shown: rate one-half ($R\frac{1}{2}$), rate $\frac{7}{8}$, and a triple Golay code. The coding adds additional bits to the message, thereby increasing the bandwidth and the noise bandwidth. The coding permits the detection and correction of errors. This process is more efficient than the extra noise al-

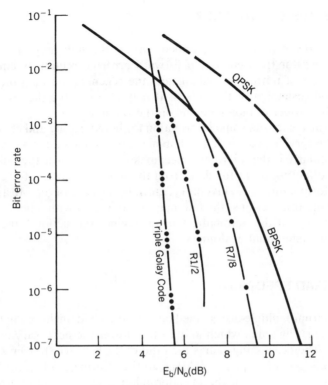

Figure 13.13. Error performance.

lowed in by the wider bandwidth and is particularly useful for this type of telecommunications system, where the final link is so power limited.

The triple Golay code, will operate with E_b/N_0 ratios of approximately 5 dB. These may be used in applications where it is critical that the message gets through in spite of a fading signal strength, for example. The triple Golays may also be used for addresses while the actual information part of the message may use a different type of coding. This is done on the philosophy that it is more important to get a message to the right address (even if the message might have some errors) than have a message arrive at the wrong address.

The forward error-correction methods are vital to increase the number of channels that may be accommodated in a transponder (particularly, the final link). The type of coding may be changed as the signal moves from one link to another. As we have seen earlier, the final link is power (not bandwidth) limited, and therefore the coding may be used while reducing the power requirements by 2–3 dB, which is immediately translated into a substantial increase in capacity (up to 100%). Since the final link is not usually bandwidth limited, there is spectrum for these extra (coding) bits.

13.7   THE HUB CONTROLLER

In addition to the functions described earlier, the hub may also act as a digital regenerator (particularly, if forward error correction is employed). As in the case of normal digital signals, the regeneration may include the retiming and reshaping of the pulses. It may also involve the switching of traffic to the proper outgoing beams on the satellite.

The hub also extracts some information for bookkeeping and charging the users for the service provided. The hub may also be located at a central data base (particularly if the microterminals make inquiries or if the hub acts as a central collecting point for data from the microterminals: for example, weather information, sales, inventory, etc.). In the case of remote automatic teller machine transactions, the hub may be colocated with a bank's central computer to speed the automatic teller machine transactions, thereby accommodating additional customers.

13.8   SPREAD SPECTRUM

Spread-spectrum multiple-access methods using code-domain multiple access provide processing gains which may further lower the necessary carrier-to-noise ratio. The processing gain is the ratio of the bandwidth over which the modulation is spread, typically 5 MHz, to the information bandwidth. This is the same as the ratio of the pseudorandom code bit rate (chips) to the information rate.

Spread-spectrum multiple access (SSMA) systems take a relatively narrow band (low bit rate) signal and spread it over a much wider-than-normal bandwidth through the use of the code domain. Since each user has a unique (nonoverlapping) code, a multiple access method is provided, hence the term SSMA/CDMA. The spreading is often introduced by direct-sequence PSK modulation.

There are a limited number of codes that are orthogonal for simultaneous multiple access use, and synchronization may take a few minutes. The spectral efficiency of SSMA is low but this is more than offset by the convenience and economics of the small terminals. The use of SSMA/CDMA reduces interference to other users.

SSMA had the disadvantage of being more complex until the arrival of custom large-scale integrated circuits. Among its commercial advantages are the ability to suppress interference from other SSMA users and withstand a lot of terrestrial (point-to-point telephony) interference and the self security of unique codes. Even solar noise can be accommodated in the SSMA microterminals.

The low required power levels pose a low-interference threat to other systems, and therefore the microterminals are easier to license and operate. The major problems of interference and low satellite eirp at C-band are not present at K_u-band, where unlimited power and a satellite exclusive band are available.

In late 1985, Pelton and Lemus described the Intelsat version of spread spectrum called Intelnet [1]. Table 13.11 compares the domestic and international systems. The operational Intelnet service parameters may vary from those shown.

TABLE 13.11   Typical Commercial Spread-Spectrum Systems

	Domestic	International	Unit
Small Terminal			
Operating frequencies			
Earth to space	5.925–6.425	14.0–14.5	GHz
Space to earth	3.700–4.200	10.7–12.5	GHz
Antenna			
Receive-only diameter	0.6 (2)	—	m (ft)
Transmit/receive diameter	1.2 (4)	1.2 (4)	m (ft)
Receive gain	25.5 and 31.5	40.8	dBi
Transmit gain	35.0	42.5	dBi
Transmitter rating	1	1.3	Watt
EIRP capability (after line losses)	34.0	42.4	dBW
Receiving system noise temperature	130	300	K
Figure of merit (G/T)	+4.4 and +10.4	+16.0	dBi/K
Price per terminal (commercial)			
Receive-only	$2000		
Transmit/receive	$6000	?	
Quantity installed to mid 1987	30,000	100 (estimated)	
Installation rate	100	?	per week
Information rate (baseband)	19.200 (42.8)	9.600 (39.8)	kb/s (dBb/s)
Chip rate	About 2.5 (64)	2.4576 (64)	Mb/s (dBb/s)
Bandwidth	5	5	MHz
Processing gain	21.2	24.2	dB
Hub Terminal			
Operating frequencies			
Earth to space	5.925–6.425	14.0–14.5	GHz
Space to earth	3.700–4.200	10.7–12.5	GHz
Antenna			
Receive/transmit diameter (typical)	10 (32)	10 (32)	m (ft)
Receive gain	49.8	59.4	dBi
Transmit gain	53.4	60.9	dBi
Receiving system noise temperature	100	350	K
Figure of merit (G/T)	29.8	34	dBi/K

TABLE 13.12 Spread-Spectrum Systems: Uplink from Microterminal to Satellite

Parameter	Domestic	International	Unit
Uplink frequency range	5.925–6.425	14.0–14.5	GHz
Transmitter power level	1	1.5	W
Spread bandwidth (2.5 Mchips)	5	5	MHz
Power to antenna	−67	−65.4	dBW/Hz
Antenna size	1.2	1.2	m
Antenna gain	+35	+42.5	dBi
Beamwidth	2.9	1.2	deg
EIRP density	−32	−23	dBW/Hz
Illumination level density at the satellite	−195	−186	dBW/m^{2}Hz
Satellite (typical)	*Westar IV*	*Intelsat V*	
Satellite antenna on axis gain	+29	+39	dBi
Satellite system noise temperature	1050	1000	K
	30.2	30	dBK
Satellite G/T	−1.2	+9	dBi/K
Off-axis allowance	2	2	dB
Effective G/T	−3.2	+7	dBi/K
Uplink C/N$_0$	−7	+5	dB

TABLE 13.13 Spread-Spectrum Systems: Downlink from Satellite to Hub

Parameter	Domestic	International	Unit
Downlink frequency	3.7–4.2	10.7–12.5	GHz
Satellite power amplifier	5	7.5	W
Estimated power density	−76.2	−105	dBW/Hz
Transmitter antenna gain	+29	+39	dBi
EIRP density	−48	−66	dBW/Hz
Bandwidth (2.5 Mchips)	5	5	MHz
EIRP/carrier	+19	+1.3	dBW
Off-axis loss	2	2	dB
Illumination level density	−213	−231	dBW/m^2/Hz
Earth station antenna diameter	10	10	m
Earth station antenna gain	+49.8	+59.4	dBi
Beamwidth	0.5	0.17	deg
Earth station system noise temp.	100	350	K
	20.0	25.4	dBK
Earth station G/T	+29.8	+34.0	dBi/K
Downlink C/N$_0$	+12.0	−11.5	dB
Uplink C/N$_0$ from microterminal	−7	+5	dB
Total C/N$_0$	−7.1	−11.6	dB
Information rate	19,200	9,600	b/s
Chip rate	2.5	2.5	Mb/s
Ratio of rates (proc. gain)	21.3	24.1	dB
C/N in information bandwidth	+14.2	+12.5	dB

TABLE 13.14 Spread-Spectrum Systems: Uplink from Hub to Satellite

Parameter	Domestic	International	Unit
Uplink frequency range	5.925–6.425	14.0–14.5	GHz
Transmitter power level	0.016	0.001	W
Spread bandwidth (2.5 Mchips)	5	5	MHz
Power to antenna	−87	−99	dBW/Hz
Antenna size	10	10	m
Antenna gain	+53.4	+60.9	dBi
Beamwidth	0.35	0.15	deg
EIRP density	−33.6	−38.1	dBW/Hz
Illumination level density at the satellite	−197	−201	dBW/m²/Hz
Satellite (typical)	*Westar IV*	*Intelsat V*	
Satellite antenna on axis gain	+29	+39	dBi
Satellite system noise temp.	1050	1000	K
	30.2	30	dBK
Satellite G/T	−1.2	+9	dBi/K
Off-axis allowance	2	2	dB
Effective G/T	−3.2	+7	dBi/K
Uplink C/N₀	−8.7	−10.0	dB
Downlink frequency	3.7–4.2	10.7–12.5	GHz
Satellite power amplifier	5	7.5	W
Estimated power density[a]	−76.5	−86.2	dBW/Hz
Tramsmitter antenna gain	+29	+39	dBi
EIRP density	−48	−47.2	dBW/Hz
Bandwidth (2.5 Mchips)	5	5	MHz
EIRP/carrier	+19	+19.8	dBW
Off-axis loss	2	2	dB
Illumination level	−213	−212.2	dBW/m²
Earth station antenna diameter	1.2	1.2	m
Earth station antenna gain	+31.5	+40.8	dBi
Beamwidth	4.4	1.5	deg
Earth station system noise temp.	130	300	K
	21.0	24.8	dBK
Earth station G/T	+10.4	+16.0	dBi/K
Downlink C/N₀	−7.4	10.7	dB
Uplink C/N₀ from microterminal	−6.9	−8.8	dB
Total C/N₀	−10.1	−12.9	dB
Information rate	19,200	9,600	b/s
Chip rate	2.5	2.5	Mb/s
Ratio of rates (processing gain)	21.3	24.1	dB
C/N in information bandwidth	+11.2	+11.2	dB

[a]Includes losses between power amplifier output and transmit antenna input and multicarrier backoff allowances.

Tables 13.12 through 13.14 trace the flow of the communications from the microterminal to the hub and on to the distant microterminal for two-way service. As may be noted in the above-mentioned tables, negative carrier-to-noise density ratios are frequent in spread-spectrum systems. The processing gain makes up for the lack of signal strength and yields a final value (see the last line of Table 13.14) of approximately 11 dB, which is sufficient to provide a high-quality signal.

13.9   ALOHA MULTIPLE-ACCESS SYSTEMS

The ALOHA systems operate on a contention basis and are often used to signal that the microterminal has traffic for the hub. Figure 13.14 shows the performance of several ALOHA systems. The vertical scale (utilization) shows the fraction of the total capacity of the transponder that can be utilized. The horizontal axis shows the traffic flowing through the communications medium. Individual earth stations may transmit at any time to the hub through a common channel. Other stations contend for the attention of the hub.

As long as there are only a few stations or when the traffic is low, the likelihood of a collision between the request packets is small. This is illustrated in the lower left side of the figure, where the traffic follows the theoretical limit for a fully transparent transponder. As the channel gets busier (more stations requesting the attention of the hub), contention begins to take place and packets start to collide.

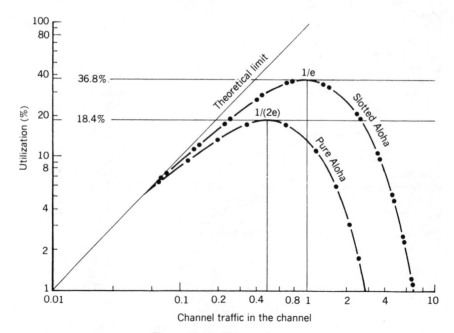

Figure 13.14. Aloha performance.

Note that the curve for pure ALOHA begins to bend toward the right. Packets destroyed by collision must be retransmitted (each at random times to avoid a new collision of these packets). This adds further to the congestion and the number of collisions. Finally, when the utilization reaches 18.4%, the collisions become so frequent and the retransmissions so common that the system begins to collapse. Eventually, the throughput drops toward zero (right side of the curve). If the traffic subsides or some packets are withdrawn, the network may suddenly find itself on the left side of the hump and the congestion is unlocked.

As the traffic builds once again, the system can become locked. These systems have a bistable characteristic as far as throughput is concerned. The slotted ALOHA divides the time domain into slots equal to a single packet transmission time and as such begins to look like TDMA without assignments. Packets therefore cannot overlap and either there is contention for a time slot or no contention. Eventually, this contention leads to the same type of flip-flop or bistable instability as the pure ALOHA, but the channel utilization is doubled. Slotted ALOHA is used in some of the microterminal systems.

13.10 TRANSPONDER REDUNDANCY

In Figure 13.4 it is noted that there are only two types of active equipment in a transponder: (1) the receivers, and (2) the amplifiers. It is the active equipment that is most subject to failure, and therefore redundancy is provided for that equipment. When a failure is detected, the redundant equipment is switched in by ground command to replace the failed equipment.

13.10.1 Receiver Redundancy

There are four receivers on the satellite shown in Figure 13.4. Only two are required for full operation. Any particular sequence of two failures may be postulated and there will be an appropriate restoration sequence.

The following scenario illustrates one failure sequence and subsequent restoration actions using the redundant receivers and assuming that the switches always work.

- Receivers 2 and 3 operational.
  Receivers 1 and 4 in standby.
- Receiver 2 fails.
  Receiver 1 is switched in to replace receiver 2.
  Receivers 1 and 3 oprational.
  Receiver 4 in standby.
- Receiver 1 fails.
  Receiver 3 is switched to replace receiver 1 and receiver 4 is switched to replace 3.
  Receivers 3 and 4 operational.

13.10.2 Transponder Amplifier Redundancy

The transponder amplifier is an assembly that may consist of:

1. A driver limiter amplifier
2. A variable (switchable) attentuator
3. A power amplifier (TWTA or SSPA)
4. An electric power conditioner

In a typical U.S. K_u-band satellite, there are 16 operational transponder channels and six extra amplifiers which serve as standby replacements for any failed units. The numbering plans for amplifiers and transponder channels may be independent since amplifiers are switched among transponder channels to provide restoration. In Figure 13.4, it may be noticed that the transponder channel number differs from the amplifier number.

The transponder amplifiers are arranged in a 22-for-16 ring redundant configuration (Figure 13.4). This means that there are 6 spare transponder amplifiers interspersed among the 16 operational amplifiers, that each amplifier can be replaced by another amplifier adjacent to it in either direction, and that every amplifier has two adjacent amplifiers (the amplifiers are connected in a ring).

The GSTAR configuration is shown in Figure 13.15, where:

- PA 7 and PA 9 are operational 27-W high-power amplifiers.
- PA 8 is the 27-W high-power-amplifier spare.

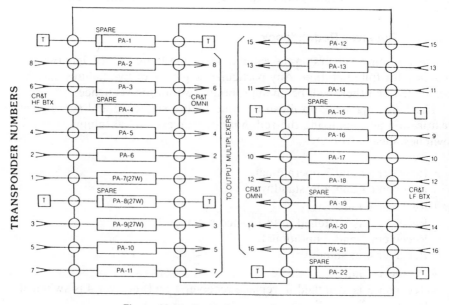

Figure 13.15. Redundancy configuration.

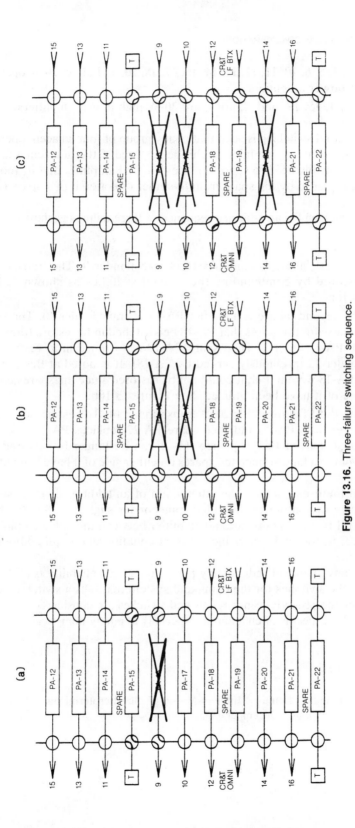

Figure 13.16. Three-failure switching sequence.

191

- PA 2, 3, 5, 6, 10, 11, 12, 13, 14, 16, 17, 18, 20, and 21 are 20-W operational high-power amplifiers.
- PA 1, 4, 15, 19, and 22 are spare 20-W high-power amplifiers.

The restoral switching is such that in any group of four adjacent operational transponders, any three failures can be restored if the adjacent spare amplifiers are available. If a spare replaces a failed amplifier and it, too, fails, it can be replaced by the spare from the other direction if that spare is available.

The switching for an arbitrary sequence of three failures is shown in Figure 13.16.

1. The first failure is amplifier 16 of transponder 9. The transponder is restored by commanding the coaxial switches as shown in Figure 13.16a.
2. The second failure is amplifier PA-17 of transponder 10. The restoral process first requires that PA-18 be switched in to restore transponder 10 (PA-17 failure). Then spare PA-19 is switched in to restore transponder 12 (previously served by PA-18). It is noted at this point that if PA-18 were to fail, it could not be restored since the spares on either side of it have been utilized. See Figure 13.16b.
3. The third failure is PA-20 of transponder 14. To restore service, PA-21 replaces PA-20 and then spare PA-22 replaces PA-21. Again it is noted that for this sequence PA-21 can no longer be restored in the event of a failure since the spares on either side of it have been utilized.

When there is a failure, the attenuation of an additional set of switches will reduce the output of the failed transponder slightly (a few tenths of a decible.) If there is an operational amplifier between the failed amplifier and the spare, it, too, will suffer increased attenuation due to an additional set of switches.

As in any switched redundancy plan, the actual reliability is affected severely if the switches (or the command system that tells a switch to switch) fail to operate properly. Since relays and switches are considered to be highly reliable, this aspect is often ignored, especially in purely numerical exercises.

REFERENCE

1. J. Pelton and R. Lemus: "Intelnet-March of the Microterminals," *Satellite and Space Technology*, December 1985.

14

SPACE SEGMENT REQUIREMENTS

14.1 CAPACITY REQUIRED FOR A MICROTERMINAL PORT

Figure 7.5 showed a projection of the telephony market shares for voice and data for the overall U.S. telecommunications marketplace. In the business portion of this market, digital data are growing even more rapidly than shown. In many instances, modems may be connected to telephone lines in the office. These may be turned on at 8:30 A.M. and remain on the telephone line until 5:00 P.M. The amount of information being transmitted at any time may be low or nonexistent (such as at lunch). From a telecommunications standpoint, the line is still active. This substantially changes the typical central office and PBX holding patterns, which are based on telephony. This adds an unexpected strain on existing local telephone offices. This is a complication for telephone networks that were designed primarily for voice services. It is uneconomical and unwise to use a satellite system in this manner.

One way to reduce the cost and better utilize existing facilities is to use packet switching. A connection to the host node at a local facility can be established and maintained. The packet switch, however, sends to another node only when it has information to transmit. Thus even though the local facilities may seem to be tied up, the long-distance facilities may be time shared by the packet switch.

A satellite network can be operated in several manners. One method is to place a packet switch at the microterminal (see Figure 14.1). The microterminal would transmit a burst of data only when it has information coming into the node from one of its sources. At other times it would be off the air and therefore could allow other microterminals to make use of the space segment.

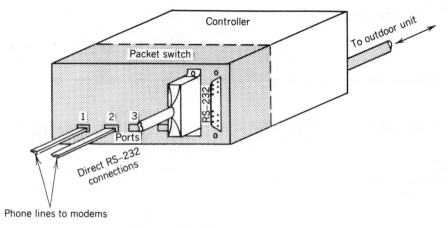

Figure 14.1. Multiport microterminal.

Please note that the microterminal may have multiple data ports. These are connected to various personal computers, facsimile machines, point-of-sale devices, monitors, and so on. In general, a microterminal with four ports connected should have twice as much traffic as a two-port microterminal. We are therefore more concerned about the number of ports connected than simply microterminals.

The other approach is to have the station on the air at all times, whether there is traffic or not. This is wasteful use of both radio spectrum and money. It has the advantage of instantaneous access to the satellite, whereas in the time-shared case, a connection must be established between the small terminal and the hub before telecommunications can commence. If the satellite channel is already busy (with another station's traffic), the local station must wait its turn before it can transmit.

In the shared space segment situation, the number of earth stations per space segment depends very much on the amount of time each station is "off hook." The problem is very similar to a large office building with several hundred telephones and a private branch exchange (PBX) switchboard. The question arises as to how many outgoing trunks to the local telephone central office should be provided. This is normally solved on the basis of Erlangs. This, in turn, is determined by the calling patterns, holding times, and the probability that all lines will be busy.

In the case of some networks (such as credit verification) the amount of time each station is on the air is relatively small. For example, it may ask the question: "Is credit card 1234567AB valid for a $21.23 purchase?" It may be several minutes before another question is asked by the microterminal. In this situation, 35 stations might be accommodated per space segment increment. As long as the traffic volume is low, a simple slotted ALOHA system can be used with a net throughput efficiency of up to 36%. (See Section 13.9.) As the traffic grows, either the ALOHA system could be replaced with something more elaborate (thereby restoring the throughput) or adding more space segment (thereby retaining the simplicity of the ALOHA system).

14.2 INBOUND LIMITATIONS (MICROTERMINAL TO HUB)

The capacity in the small terminal-to-hub direction is established on bandwidth basis. This is because the individual small transmitters do not have sufficient power to operate the satellite near saturation. If it is assumed that these are 56-kilobit links that can be accommodated in increments of 45 kHz each, Figure 14.2 shows the capacity as a function of the transponder bandwidth (see also Chapter 13).

14.3 OUTBOUND LIMITATIONS (HUB TO MICROTERMINAL)

Because the microterminals are both aperture (gain) and noise limited, they need as much power as possible from the satellite; thus the hub-to-microterminal link is power limited. Figure 14.3 shows the transponder capacity as a function of EIRP and microterminal size.

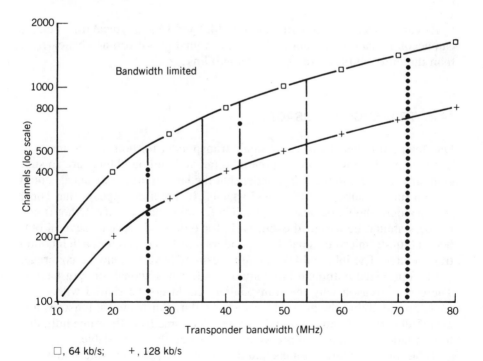

□, 64 kb/s;   +, 128 kb/s

Figure 14.2. Microterminal-to-hub capacity.

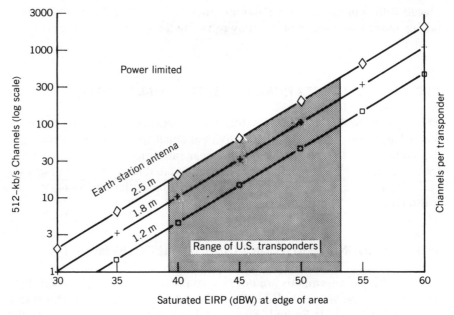

Figure 14.3. Hub-to-microterminal capacity.

As can be seen by comparing Figures 14.2 and 14.3, a typical transponder can accommodate more inbound (small terminal-to-hub) 56-64 kb/s carriers than the outbound (hub-to-small terminal) link.

14.4   TRANSPONDER USAGES

One temptation may be to use a single transponder for both directions. Unfortunately, the power levels of the inbound and outbound links are so substantially different in the transponder that the large (outbound) signals could swamp (power suppress) the small signals. It therefore appears that some transponders should be designated solely for inbound capacity. Other transponders should be assigned exclusively for outbound service. Several outbound (hub-to-microterminal) transponders may be needed for each inbound transponder. The inbound transponders usually have nationwide coverage, but an optimized outbound link may use national, regional, or spot beams. There is no reason why the transponders could not be shared with other users provided that the power levels are similar. For technical reasons, it is best if all carriers in an outbound transponder come from the same hub, due to rain fades and signal suppression. This is not always possible.

Table 14.1 compares uplinks and downlinks. If 1.2-m microterminals are

TABLE 14.1 Transponder Capacity Example[a]

Antenna	Channels	Channel Bit Rate (kb/s)	Total Channels at 64 kb/s	Channels Ratio to 1200	Reference Table
		Inbound			
9 m	1200	64	1200	—	13.3
		Outbound (Conventional TWTA, 4-dB Backoff)			
1.2 m	4	512	32	1:37.5	—
1.8 m	8	512	64	1:18.8	13.5
2.5 m	17	512	136	1:8.8	—
		Outbound (Linear Transponder, 2-dB Backoff)			
1.2 m	6	512	48	1:25	13.10
1.8 m	14	512	112	1:10.7	13.10
2.5 m	27	512	217	1:5.5	—

[a]Transponder assumptions: 45 dBW at center, 54 MHz; earth station LNA: 225 K; rain-loss accommodation: 3 dB.

used, there must be 25 linearized outbound transponders for every inbound transponder. A conventional TWTA has an even worse inbound/outbound ratio (37.5:1).

At the present time, the space segment for a 1.8-m station represents less than one-half the monthly costs (see Table 9.4). Eventually, as the available space segment is consumed, its price will rise and larger (2.5-m) antennas may be introduced to double the number of channels per space segment dollar. The 2.5-m antennas may be used to retrofit existing 1.2- and 1.8-m stations.

14.5 ASYMMETRICAL TRAFFIC

In some applications, many of the inbound data inquiries are short and the outbound answers are long. The amount of asymmetry and its direction depend of the use of the network. In the credit-card verification situation, there is more inbound information than outbound (which may be almost a simple "yes" or "no"). For inquiries into large data bases (such as for literature searches), the remote station may make a simple request "search and find all data on locomotive wheels between 1907 and 1927." The response may take far more space segment capacity than the question, particularly if extensive abstracts or entire documents are returned to the requester via the satellite link.

14.6   TRAFFIC PER SERVICE

14.6.1   Library Searches

In this service the user is seeking information from a remote data base library (e.g., Dialog). A typical search takes 20 minutes on line. To log onto the service and pose the question properly takes about 2500 alphanumeric characters. At 8 bits per character (ASCII code with parity checks), 20,000 bits are needed. The overhead bits (addresses, etc.) are about 10% more. Therefore, 22,000 bits are transmitted from the microterminal for this port (see Table 14.2). If the average usuable transmission bit rate of the microterminal during a burst is 36 kb/s (see below), the 22,000 bits can be accommodated in a series of bursts totaling about 0.6 s for this input port.

Rate ¾ forward error correction coding is used. This adds bits.

To keep bursts from overlapping, a generous (25%) guard time has been assumed. This reduces the effective throughput from 64 kb/s per channel to 36 kb/s per channel. Due to the packet switch, the 22,000 bits use the satellite only when needed. The effective average transmission bit rate for the port during the 20-minute session is 18.3 b/s. After R¾ encoding this rises to 24.4 b/s.

If the transponder channel can accommodate 36 kb/s (Table 14.2) and the per port average transmission rate is 18.3 b/s, each channel can carry up to 1,967 simultaneous library searches at the busy hour in the microterminal-to-hub direction.

Each transponder has many (e.g., 600) channels. The return link (hub to microterminal) is, on average, much more active. Approximately 30,000 characters are transmitted back to the microterminal port (see Table 14.3).

TABLE 14.2   Library Search Bit Rates: Microterminal to Hub

Line		Unit
1 Characters	2,500	characters
2 Bits per character	8	bits
3 Bits (line 1 × line 2)	20,000	bits
4 Overhead (10%)	2,000	bits
5 Bits (line 3 + line 4)	22,000	bits
6 Transmission rate	36,000	b/s[a]
7 Total burst time (line 5/line 6)	0.61	s
8 In-use time	20	min
9 In-use time (line 8 × 60)	1,200	s
10 Average bit rate (line 5/line 9)	18.3	b/s per port
11 After R¾	24.4	b/s per port
[a](a) Bit rate in burst	64,000	b/s
(b) Rate ¾ encoding	16,000	b/s
(c) Information rate (line a - line b)	48,000	b/s
(d) Guard time between bursts	25	%
(e) Actual throughput		
[line c × (100% - line d)]	36,000	b/s

TABLE 14.3 Library Search Bit Rates: Hub to Microterminal

Line		Unit
1 Pages sent	10	pages
2 Lines per page	50	lines
3 Characters per line	60	characters
4 Total characters (line 1 × line 2 × line 3)	30,000	characters
5 Bits per character	8	bits (ASCII)
6 Total bits (line 4 × line 5)	240,000	bits
7 Overhead	10	%
8 Line 6 × line 7	24,000	bits
9 Subtotal (line 6 + line 8)	264,000	bits
10 Transmission rate (see Table 14.2)	36,000	b/s
11 Total burst time (line 9/line 10)	7.33	s
12 In-use time	20	min
13 Line 12 × 60	1,200	s
14 Average bit rate (line 9/line 13)	220	b/s per port
15 After R¾	293	b/s per port

Since the session is spread over 20 minutes, the average bit rate is only 220 b/s if no encoding is done and about 300 b/s after R¾ encoding. The average bit rate must be used to determine the demands on the space segment and the hub.

The 36-kb/s effective rate per channel will support about 163 library searches in the hub-to-microterminal direction. Note that the traffic is asymmetric (1,967 versus 163 simultaneous uses per channel).

14.6.2 Routine Data Retrieval

This is a much simpler task than the entire library search. Here a specific data base is queried (e.g., insurance rates for various types of cars with certain options). The distant data base may also have a computer to determine the insurance rate based on the car type, year, location garaged, driver history, age, sex, and so on. A complete policy calculation may be done by the computer for the branch office requesting the data.

The data sent to the hub may be more extensive (the equivalent of one typed page, 3000 characters, or 26,400 bits). The response may be 3 pages (about 80,000 bits). If the session lasts 5 minutes (300 seconds), the average bit rates are 88 and 267 bits per second (or 117 and 356 bits per second after R¾ coding). A single channel can accommodate 409 users in this class between the microterminal and the hub. Only 135 users can be supported in the other direction.

14.6.3 Voice at 32 kb/s

In this case, the 32 kb/s already includes overhead FEC, and so on, supplied by the voice coder. The 64-kb/s channels in a transponder can accommodate two 32-kb/s voice channels from the same microterminal. The simplest case is for both channels to go to the same destination (two private lines). Since the hub is present, it may be possible to do switching to permit a private "dial-up" network service.

If only one 32-kb/s voice channel is present, the remaining capacity may be used for other digital services. Other voice rates (e.g., 16 kb/s and lower) are also possible.

Unlike data or record services, voice normally requires a full-time (or the equivalent) connection from each end. This ties up the space segment. Voice activity ("push to talk") may be added to reduce the space segment demands to one-half but contention for available circuits will arise.

14.6.4 Credit Verification

As explained earlier, this service requires relatively little capacity. Table 14.4 itemizes the elements in the microterminal-to-hub path. If the maximum number is used (1333 after FEC coding) and the transponder channels are 64 kb/s, up to 48 queries can be made simultaneously. After allowing a 25% guard time, this number becomes 36 per transponder channel.

The return link is detailed in Table 14.5. Between 387 and 1139 bits may

TABLE 14.4   Credit Verification: Microterminal to Hub

Line	Minimum[a]	Maximum[a]	Unit
1 Store code	10	34	characters
2 Checksum	3	10	characters
3 Separator	2	2	characters
4 Cashier station number	3	5	characters
5 Checksum	1	2	characters
6 Separator	2	2	characters
7 Credit-card number	16	20	characters
8 Checksum	5	6	characters
9 Separator	2	2	characters
10 Amount to be charged ($XXXX.XX)	8	8	characters
11 Checksum	3	3	characters
12 Separator	2	2	characters
13 Card type	5	8	characters
14 Completion code	5	5	characters
15 Subtotal	67	109	characters
16 Bits per character (ASCII)	8	8	
17 Bits (line 15 × line 16)	536	872	bits
18 Overhead (10% of line 17)	54	87	bits
19 Total (rounded)	600	1000	bits
20 R¾ FEC bits	200	333	bits
21 Grand total (line 19 + line 20)	800	1333	bits

[a]Based on an examination of American Express and Diners Club receipts.

TABLE 14.5 Credit Verification: Hub to Microterminal

Line		Minimum[a]	Maximum[a]	Unit
1	Store code	10	34	characters
2	Checksum	3	10	characters
3	Separator	2	2	characters
4	Cashier station number	3	5	characters
5	Checksum	1	2	characters
6	Separator	2	2	characters
7	Approval code	0	6	characters
8	Checksum	0	2	characters
9	Separator	0	2	characters
10	Message	0	20	characters
11	Separator	2	2	characters
12	Maximum credit limit ($XXXX)	5	5	characters
13	Completion code	5	5	characters
14	Subtotal	33	97	characters
15	Bits per character (ASCII)	8	8	
16	Bits (line 14 × line 15)	264	776	bits
17	Overhead (10% of line 16)	26	78	bits
18	Total (rounded)	290	854	bits
19	R¾ FEC bits	97	285	bits
20	Grand total (line 18 + line 19)	387	1139	bits per burst

[a]Based on an examination of American Express and Diners Club receipts.

be required (the length depends on the condition of the credit). At 64 kb/s per channel 42 to 124 verifications can be accomplished assuming 25% guard times.

14.6.5 Sales Records, Inventory Control, and so on

During the peak sales season (retail stores may make 50% of their sales in November and December) it is vital to keep track of the stock. Many stores have automated cash registers and hand-held scanning pens that read the bar codes and price tags. These data are accumulated by the store. In some cases the data are read out in the store for ordering and sales planning. In other cases, they are recorded (on disk or tape) and hand carried to a central location. In some cases the data are read out into a modem that is connected to a telephone line to the central location.

If the inventory system has been computerized, each sale may be subtracted from stock. If the rate of change of each item in stock is monitored, the fast-selling items (and those that are not selling) may be found. The effectiveness of a sales promotion may be determined.

A microterminal may be used for the link between the store and the central location. Several types of transmissions are possible, as indicated in Table 14.6.

TABLE 14.6 Sales Records and Inventory Control: Microterminal to Hub

Service	Sales Check		Inventory Control		New Prices[a]
When	Anytime	Evening	Anytime	Evening	Anytime
Stimulus	Remote[b]	Local clock or polled	Remote[b]	Local clock or polled	Remote[b]
Readout type	Summary or specific department	Complete	Specific stock item or group	Complete	n/a
Pages printed	5	15[c]	2	35[c]	n/a
Bits w/o FEC[d]	132,000	396,000	52,800	924,000	200[e]
R¾ FEC bits	44,000	132,000	17,600	308,000	67
Total bits	176,000	528,000	70,400	1,232,000	267
Readout (s)					
at 64 kb/s	2.75	8.25	1.10	19.25	0.004
Query rate					
(per day)	3	1	25	1	2
(busy hour)	0.67[f]	0	5.4[f]	0	0.5[f]
Average busy hour bits per microterminal	117,920	0	380,160	0	134
Seconds per busy hour per microterminal	1.84	0	5.94	0	0.002
25% guard time	0.46	—	1.49	—	0.0005
Total seconds	2.30	—	7.43	—	0.0025
Seconds in busy hour	3,600	—	3,600	—	3,600
Users per channel	1,565	—	485	—	Many

[a]n/a, not applicable.
[b]The central location (e.g., headquarters or the district office).
[c]Or equivalent to the central computer.
[d]Based on 26,400 bits per page.
[e]Acknowledgment bits.
[f]Based on 300 stores. Value shown is per microterminal port.

In the first data column, the remote office may check the performance of any department (or the entire store) at any time (upon request). It is assumed that half of all the inquiries occur during the busy hours.

The second data column is a scheduled transmission (usually at night after the store has closed its books). The transmission may be made according to the time of day or in response to being polled by the central office. This transmission covers all departments.

The third and fourth data columns are similar to the first two except it is for inventory control. The final column is the acknowledgment channel for price change lists and administrative notices sent from the hub.

Table 14.7 shows the return link (to the microterminal). The inquiries take little time. The price and policy transmissions take more capacity.

TABLE 14.7   Sales Records and Inventory Control: Hub to Microterminal[a]

Service:	Sales Check		Inventory Control		New Prices
	Hub select[b]	Evening	Hub select[b]	Evening	Hub select[b]
To:	One store	One store	One store	One store	All stores
Transmission	Brief inquiry	Poll	Brief inquiry	Poll	Longlist of new prices or policies
Pages	n/a	n/a	n/a	n/a	25
Bits w/o FEC	4,000	1,000	4,000	1,000	660,000
R¾ FEC bits	1,333	333	1,333	333	220,000
Total bits	5,333	1,333	5,333	1,333	880,000
Transmit time at 64 kb/s (s)	0.08	0.02	0.08	0.02	13.8
Transmission per hour	45,000	180,000	45,000	180,000	260
After 25% guard time	33,750	135,000	33,750	135,000	200

[a]n/a, not applicable.
[b]It is assumed that the central office is at the hub (or communicates via the hub).

14.6.6   Teleconferencing

Although 64-kb/s teleconferencing equipment is available, many users prefer the 128-kb/s (or higher) bit rate. The 128 kb/s can be obtained through allocation of some wider (e.g., 90 kHz) channels or by ganging two adjacent 64-kb/s (45-kHz) channels. The ISDN 2B + D operates at 144 kb/s. In any case, each direction of the teleconference requires full-time (non-time-shared) service.

14.6.7   Group IV Facsimile

The 64-kb/s requirement for this equipment indicates that each facsimile microterminal port will fully occupy a channel during its transmission. If multiple machines are connected and there is only one RF channel available, the facsimile pages will be queued up. Since pages may be both sent and received, the traffic is symmetrical for single address messages.

14.7   MICROTERMINALS PER TRANSPONDER

The answer to the question of how many microterminal ports can be accommodated by a transponder is a function of many variables. As indicated in the previous sections, the capacity requirements for the seven different types of traffic varies substantially. We also expect the number of terminals

TABLE 14.8   Service Traffic Capabilities

		Ports per 64-kb/s Channel[a]	
Application	Use Time (min)	Microterminal to Hub	Hub to Microterminal
Library searches	20	1967	163
Routine data retrieval	5	409	135
Voice at 32 kb/s	5	2	2
Credit verification	1	36	42–124
Sales records, inventory, etc.	1	485–1565	200–33,750
Teleconferencing	40	0.5	0.5
Group IV facsimile	0.5	1	1

[a]36 kb/s after R¾ FEC and 25% guard time allowances where applicable.

to be substantially different among the services, with credit verification probably being the largest near term potential user of microterminals.

Table 14.8 shows the service traffic capabilities of each of the seven applications. Both the inbound and outbound capabilities are given and it should be noted that, in general, the traffic is asymmetric.

Table 14.9 shows the microterminals per transponder in the microterminal-to-hub direction. Figures 14.4 and 14.5 are derived from this table. They show that although credit verification accounts for the largest single group of microterminal ports, this service requires only about 5% of the inbound transponder capacity. This is because the bursts are so short. The largest single user of transponder capacity is teleconferencing, although it is only used by 5% of all the ports in service.

To test the sensitivity of the results to the assumptions, we created a second assumption set. This is shown in Table 14.10. This assumption set has less facsimile, teleconferencing, and voice than the first set. The number of ports that can be accommodated per transponder is nearly 2½ times as

TABLE 14.9   Microterminals per transponder: Microterminal to Hub Assumption Set 1[a]

Application	Percent of All Ports	Ports per Transponder	64-kb/s Channels per Transponder
Library searching	3	87	0.04
Data retrieval	10	288	0.71
Voice at 32 kb/s	15	433	216.35
Credit verification	50	1442	36.06
Sales and inventory	15	433	0.72
Teleconferencing	5	144	288.46
Group IV facsimile	2	58	57.70
Total	100	2885	600 (rounded)

[a]Transponder: 54 MHz wide (600 channels of 64 kb/s). These data are shown in Figures 14.4 and 14.5.

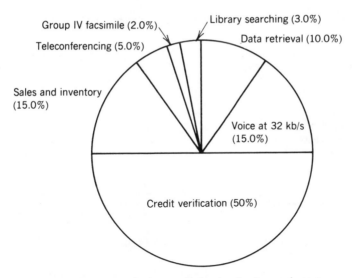

Figure 14.4. Uses of microterminals, showing types of users.

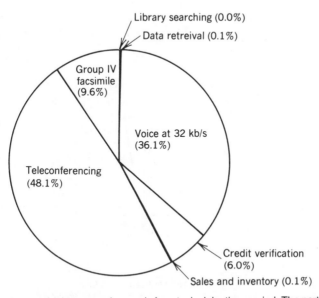

Figure 14.5. Microterminal transponder needs for a typical daytime period. The portions change at night and on weekends. In a retail store network, for example, the credit verification becomes dominant during store hours on Saturday and Sunday.

TABLE 14.10 Microterminals per Transponder: Microterminal to Hub Assumption Set 2[a]

Application	Percent of All Ports	Ports per Transponder	64-kb/s Channels per Transponder
Library searching	3	216	0.11
Data retrieval	10	721	1.76
Voice at 32 kb/s	8	577	288.43
Credit verification	50	3605	90.13
Sales and inventory	27	1947	3.24
Teleconferencing	1	72	144.21
Group IV facsimile	1	72	72.11
Total	100	7210	600 (rounded)

[a]Transponder: 54 MHz wide (600 channels of 64 kb/s). These data are shown in Figures 14.6 and 14.7.

many in the second set. A comparison of Figures 14.5 and 14.6 reveals the sensitivity to the assumptions.

The tables and figures indicate that sellers of microterminals should concentrate on credit verification, retail sales, and inventory control and be prepared to offer voice services through the microterminals. Most corporate networks using terrestrial facilities are mixture of voice and data. It is therefore not surprising that voice should appear as a candidate for the microterminals. The importance of the observations are that voice takes a substantial fraction of the space segment. Teleconferencing, another business service, has even a larger impact. From the standpoint of the marketer of the space segment, promotion of the voice and video aspects is very important.

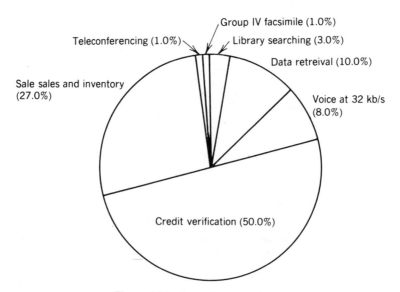

Figure 14.6. Uses of microterminals.

TABLE 14.11   Outbound Capacity: Hub to Microterminal Assumption Set 1

	Percent of Total Ports	Ports per 54 MHz	Channels in 54 MHz
Library searching	3	87	0.53
Data retrieval	10	288	2.14
Voice at 32 kb/s	15	433	216.35
Credit verification	50	1442	24.04
Sales and inventory	15	433	2.16
Teleconferencing	5	144	288.46
Group IV facsimile	2	58	57.69
Total	100	2885	591.37
Rounded	100	3000	600

Table 14.11 looks at assumption set 1 for the hub-to-microterminal link. In this case the percentages of the terminals is the same as in Table 14.9. After all, these are the same terminals, so the percentages are fixed in both directions. We have obviously made the assumption that all terminals are two-way. For this reason the number of ports per transponder is the same because of the bandwidth limitations on the inbound link predominate. The outbound link requires very slightly less channel capacity than the corresponding inbound. This is because of the asymmetry in the traffic and the overwhelming impact of voice and video services. Although the distribution of the channel capacity in the smaller users of the space segment capacity is different, the overall result is virtually the same because of the voice and video aspects.

Figure 14.7 shows the microterminal transponder needs for the hub to

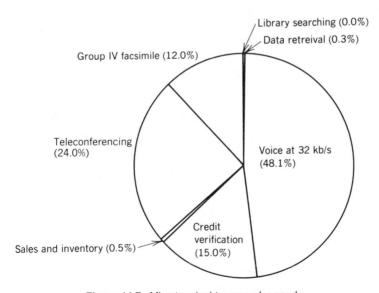

Figure 14.7. Microterminal transponder needs.

TABLE 14.12 Outbound Capacity: Hub to Microterminal Assumption Set 2

	Percent of Total Ports	Ports per 54 MHz	Channels in 54 MHz
Library searching	3	216	1.33
Data retrieval	10	721	5.34
Voice at 32 kb/s	8	577	288.43
Credit verification	50	3605	60.09
Sales and inventory	27	1947	9.73
Teleconferencing	1	72	144.21
Group IV facsimile	1	72	72.11
Total	100	7211	581.24
Rounded	100	7000	600

microterminal link using assumption set 2. In assumption set 2 (Table 14.12) this pattern is repeated (Figure 14.8).

As these two sets of assumptions have shown, one transponder will support a wide number (3000 and 7000 in these two samples) of microterminal ports in either direction. Other mixtures of services will yield different numbers. Two transponders are necessary for two-way service. Table 14.13 shows several ways of expressing this relationship.

Expressed in terms of 512-kb/s channels, each transponder must be capable of supporting 75 channels.

From Figure 14.3 it can be seen that this level of capacity in a transponder requires approximately 50 dBW from the satellite if the earth stations are

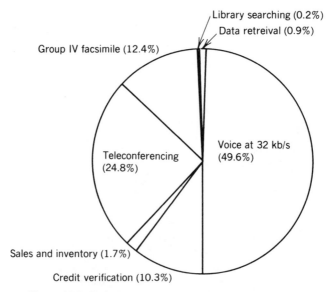

Figure 14.8. Hub-to-microterminal capacity shares.

TABLE 14.13 Transponder Capacity[a]

Assumption Set Number	Channels	One-Way Microterminal to Hub	One-Way Hub to Microterminal	Two-Way
1	In 54 MHz	3000	3000	1500
	In 64 kb/s	5	5	2.5
	At 512 kb/s		75	
2	In 54 MHz	7000	7000	3500
	In 64 kb/s	11.7	11.7	5.8
	At 512 kb/s		75	

[a]54-MHz transponders with proper power levels for the microterminal antenna sizes (see the text). All values have been rounded.

1.2 m in diameter, 47 dBW for 1.2-m stations, and about 45 dBW for 2.4-m antennas. These power levels imply either more powerful power amplifiers or spot beams. If sufficient power cannot be obtained, the traffic must be spread across more transponders to keep the power density (watts per hertz) constant, thereby increasing the number of outbound transponders necessary.

In the early years, each microterminal may have one port active. As the users become accustomed to the terminals, we expect that this will rise to 1.3 and eventually an average of 1.5 ports per microterminal. In this way, a microterminal will be to telecommunications as a minicomputer with remote work stations is to data or word processing.

Table 14.14 shows the number of transponders for a mature system. Actually, twice this theoretical number may be needed due to inefficiencies. An efficiency of 50% may be realistic because networks may lease or buy full transponders, but only use portions (the rest is in reserve for a failure, growth, etc.) *This doubles the number of transponders* (Figure 14.9). Since

TABLE 14.14 Space Segment Required[a]

Assumption Set Number	Number of Ports[b]	54-MHz Transponders Needed (Two-Way)[c]
1	100,000	67
	300,000	200
	500,000	333
	800,000	533
2	100,000	29
	300,000	86
	500,000	143
	800,000	229

[a]See also Figure 14.9.
[b]There are an average 1 to 1.5 ports per microterminal, depending on its age and application (see the text).
[c]If the hub-to-microterminal transponders do not have sufficient power, more transponders will be needed.

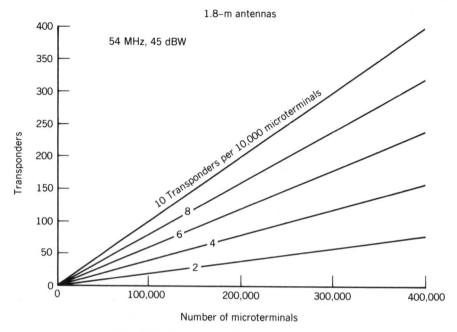

Figure 14.9. Transponders to support microterminals.

most of the initial demand is at K_u-band, this level of demand could constrain the limited supply.

If the space segment supply further tightens (and the price rises), there will be interest in increasing the microterminal antenna diameter (to 2.5 m) to double the space segment capacity (or halve the per channel space segment cost). Many of the future earth stations may be T1, for which proportionally more space segment capacity must be utilized. The ratio of the bandwidths of 1.544 Mb/s and 64 kb/s is 24:1, which indicates that in place of about 3000–7000 64-kb/s earth stations per transponder that only 125–300 T1 stations can be accommodated. This will further increase the number of transponders necessary.

14.8   SPREAD-SPECTRUM MICROTERMINALS

In SSMA, users may share the same spectrum simultaneously through the use of CDMA. When channels are added, the extra channels look like noise and the intermodulation products degrade the C/N until eventually there is difficulty in receiving the data. SSMA is spectrum inefficient with typically 10–50 carriers per transponder. A current upper limit for Intelnet I is six individual spread-spectrum carriers per 5-MHz increment. In a 36-MHz transponder, 45 such channels can be accommodated simultaneously if there is enough power.

14.9 INCREASING THE SUPPLY OF THE SPACE SEGMENT

If insufficient new satellite capacity is available, the potential market for these small terminals will saturate. When this begins to happen, the industry may react by finding more efficient ways of utilizing the spectrum in the existing satellites.

14.9.1 Signal Modifications

If there is a shortage of spectrum and satellite capacity, we believe that techniques such as bandwidth compression and advanced modulations will be employed to make better use of the resources.

14.10 LOAD SPREADING

If there is a shortage of spectrum, it will show up first at the peak busy hour (see Figure 5.8). Therefore, there may be economic incentives to send data and other services at other times of the day.

Price elasticity is defined as the percent change in demand divided by the percent change in price. In some studies conducted by AT&T, this was as high as 2.5:1. In this case, a 10% rate reduction would result in 25% more traffic during these otherwise quiet hours.

14.11 SPOT BEAMS

As indicated in Chapter 13, spot beams increase the power and thus the number of channels per transponder for a given power amplifier rating.

14.12 SSMA

Spread-spectrum multiple access is one way to accommodate the interference that could come from crowded K_u-band satellites. This might be used by new systems to coexist with existing low-power systems. This is possible, as SSMA is interference tolerant. Spread-spectrum multiple access (using weak carriers) cannot tolerate *very high* levels of interference (such as from a cofrequency television services that might accidentally appear on a SSMA transponder).

We also predict use of the principle of sharing a transponder between several services (such as television occupying part of the spectrum and most of the power of the transponder) and other services (such as SCPC occupying the rest of the bandwidth and very little of the power). Experiments of this type have been carried out under names such as TV-plus.

14.13 OPTIMIZED TRANSPONDERS

The demands on the satellite transponders in the inbound (microterminal to hub) directions are based on bandwidth, not power. Figure 14.10 shows the relationship between the total mass of a power amplifier and its RF rating. Solid-state power amplifiers start at a lower total mass, but their lower efficiency results in a higher total mass at high power.

Traveling-wave-tube amplifiers (the tubes and the high-voltage power supplies) have a large initial mass which increases with the power output. Eventually, high-efficiency TWTAs are introduced to limit the dc power requirements and therefore the total mass since the vertical scale includes a pro rata share of the spacecraft's total power subsystem.

Since the satellite-to-hub link is bandwidth (and not power) limited, it is advantageous to use a wideband transponder for this application. As long as the uplink power amplifier requirements on the microterminal are modest (below 10 W), the use of uplink spot beams in the satellite has limited appeal. Therefore, a wideband, wide-area-coverage transponder for inbound service is acceptable. There are advantages to using spot beam transponders if it is desirable to have many transmit-only microterminals with even smaller an-

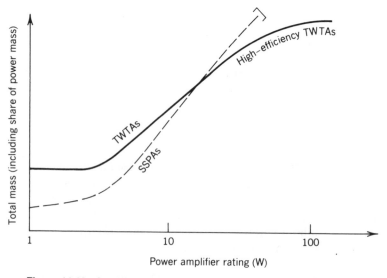

Figure 14.10. Combined equipment and power mass per transponder.

tennas, lower-powered amplifiers or a higher bit rate (e.g., T1). The smaller antenna diameter must be traded off against the increased potential for adjacent satellite uplink interference.

The transponder for the outbound (hub to microterminal) link is power limited. As the beamwidth of the outbound satellite beam (space to earth) is narrowed, the EIRP is raised, thereby permitting more traffic per dc watt, a lower-power amplifier in the satellite, or a trade-off between these two parameters.

In the early 1970s, the author [1] pointed out that a satellite composed of very small spot beams requires very little dc and RF power generation, thereby saving mass in these areas. The spacecraft becomes an antenna supported by a structure and housekeeping elements. Since the multibeam feed permits reuse of the same antenna, reflecting service for each of the beams, the total mass of the spacecraft (with many beams) may be substantially smaller than a conventional satellite.

As the number of beams is increased, the total spacecraft mass increases to a maximum and then decreases (see Figure 14.11). We do not anticipate this type of spacecraft in North America for several generations. The original designs of the Federal Express and Martin Marietta's satellites utilized many spot beams but are still on the left side of the peak.

The outbound link should have as much power as possible. Since the bandwidth utilized is relatively narrow, transponders of 36 MHz or less (and preferably in the 20-MHz range) would provide a better use of the available spectrum. At 20 MHz per transponder, at least 42 dual polarized transponders could be accommodated per satellite in 500 MHz.

Coverage may be done with spot beams whose diameter is inversely proportional to the population to help even spread the traffic across the satellite. If this is not done, the satellite saturates as soon as the busiest beam is filled.

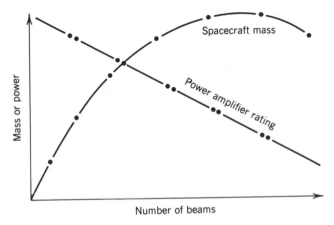

Figure 14.11. Multibeam satellite characteristics.

If the New York beam saturates, idle capacity in another spot beam (e.g., one serving Las Vegas) cannot be used to alleviate New York's problem.

Smaller beams or increased power amplifier capability is necessary into areas where rain is frequent and heavy (see Section 10.6). The alternative is to increase the earth station antenna diameter. In the United States, the area of greatest concern is along the Gulf of Mexico, including Florida, Georgia, Mississippi, Alabama, Louisiana, Tennessee, and eastern Texas. Since there are important business centers in both Florida and East Texas, beams on these areas may be advisable.

On-board signal processing, in the satellite, can further lower the cost of a microterminal, and potentially eliminate the hub station, by performing some of these functions in space. Each link (microterminal to satellite, satellite to hub, hub to satellite and satellite to microterminal) may be individually optimized through the use of on-board demodulation and remodulation.

Error control may be done on an individual link basis, thus preventing an accumulation of errors. The uplinks may arrive with variations in timing, protocols, and so on. The on-board processor can reformat, retime, and reassemble the data after performing switching and digital value-added functions. Eventually, the satellite may replace most of today's hub station's functions.

One advantage of eliminating the need to pass through the hub is that the mesh network path lengths can be halved (microterminal to satellite to microterminal, instead of microterminal to satellite to hub to satellite to microterminal). This halves the delay, which is important for voice and teleconference uses. It also cuts the spectrum needs in half.

The potential demand for transponders is very dependent on the design of the satellites and their transponders. Higher-powered beams will permit smaller and still less expensive microterminals. The use of spot beams permits more frequency senses, thus increasing the in-orbit capacity and postponing the day of saturation. As the number of beams increases, the satellite's deployed antenna diameter must grow. This could result in the need for larger launch vehicle fairings or unfurlable antennas.

14.14   NONGEOSTATIONARY ORBITS

If the geostationary orbit gets too crowded (the demand for geostationary transponders outstrips the potential supply), it will be necessary to use alternative orbits. The two principal alternatives are inclined geosynchronous orbits (narrow figure eights in which a constellation of satellites is placed in one location, and these satellites are phased so that one is always near the equator, another near the maximum inclination angle north of the orbit, and the third at the other extreme to the south. Actually, more than two satellites may be required because of the motion. If the inclination and the number of satellites is large, different earth stations may point at different portions of the figure eight and achieve isolation from the others (assuming that the antenna patterns discriminate against the other satellites). From a launch vehicle standpoint, these geosynchronous satellites have requirements very similar to the normal geostationary launchings.

Another alternative is to use the orbit typified by the Soviet Molniya series. This is an extremely elliptical orbit with a perigee in the southern hemisphere and the apogee approximately 35,000 km over 65° north latitude. If the microterminals have wide enough beamwidths, a satellite will stay within the beam for many hours before being replaced by another satellite. Since the right ascension of one system of phased satellites can be different from that of another system (orbit planes), multiple systems may be operated without mutual interference if they are sufficiently separated in space.

14.15   REVERSE FREQUENCY ALLOCATIONS

There are ways to increase the capacity of the orbit. One is to reverse the up and downlink frequencies. This frequency reversal (e.g., uplinks at 12 GHz and downlinks at 14 GHz) could nearly double the frequency resource. Caution must be observed to protect the radio astronomy frequencies near 14.5 GHz from interference; therefore, this may result in slight less than a 2:1 increase. Coordination between earth stations in systems using reverse frequency plans would be needed.

14.16   TRANSATLANTIC SATELLITES

With the minor exception of a longer slant range (approximately 0.5 dB of increased attenuation), the domestic and transatlantic satellites are very similar in their operation. Transatlantic satellites have the advantage of collecting and transporting information to small areas (as viewed by the satellite), and therefore the antenna gain can be made higher. All of the other parameters are very similar to the national situation, with the obvious exception of the frequency allocation in the various regions of the world. If the economic and political difficulties with international telecommunications are ignored, it would be expected that the same technical and marketing aspects would apply to this service.

The advent of the "market that never sleeps" (a reference to worldwide stock markets being open at all times in various cities) may further foster the use of international microterminals for at least the distribution of data and the collection of buy and sell orders.

Present international satellites (with their wide-area global, hemi, and zone beams) are unsatisfactory for this type of service. The number of K_u-band spot beams on the Intelsat V series of satellites is severely limited (to two). It is anticipated that future satellite design for the transatlantic satellite service would utilize multiple spot beams that may either be moved or (more likely) switched. Since some traffic may be addressed to multiple destinations and countries, there may be more outbound traffic than inbound in international satellite systems with microterminal applications.

REFERENCE

1. W. L. Morgan, "Communications Satellites and New Technology," *Microwave System News*, April/May 1974.

MICROTERMINAL INSURANCE ASPECTS*

15.1 RISKS AND EXPOSURES

The pointing of hundreds and thousands of antennas at a single satellite location poses unique risks due to:

Loss of a transponder(s),
Loss of an entire spacecraft,
Intentional or unintentional interference to up or down links, and
Rain outages.

Business interruption and extra expenses are risks that should be insured. "Business interruption" insurance covers financial losses resulting from an outage and loss of revenue or profits. "Extra expense" insurance covers any additional costs of doing business such as repointing earth stations or acquisition of alternative transponder capacity.

15.2 EXTRA EXPENSE INSURANCE

If it becomes necessary to move a microterminal or VSAT network to another satellite location due to transponder or satellite failure, the earth station an-

*This chapter is adapted from information prepared by Ann M. Deering of the Space Systems Group of Johnson & Higgins, 125 Broad Street, New York, NY.

tennas must be repointed. Antenna repointing is considered an extra expense risk. In the late 1980s, the estimated cost of repointing a 1000 station network could range from $250,000 to $1,000,000 and take from two weeks to three months to complete. During the transition, it may be necessary to temporarily operate duplicate transponders to keep the network alive. If a satellite is moved more than once, multiple repointings may be required.

15.3 COST OF ALTERNATE TRANSPONDER CAPACITY

As indicated previously, any additional transponder service needed during a transition in satellite location or to replace a failed transponder will be an additional expense. Selection of satellites with extensive redundancy minimizes the chance of the loss of an individual transponder (see Section 13.10). As stated in Chapters 12 and 13, many factors should be considered when selecting an alternate transponder.

15.4 BUSINESS INTERRUPTION INSURANCE

Insurance is available from insurance brokers through specialized underwriting markets. A policy can be designed to cover business interruption and/or extra expenses due to a network outage in the space segment, subject to certain exclusions such as interference and war or hostile acts. Insurance rates fluctuate almost daily and are sensitive to underwriting market conditions at the time a program is placed. Rates declined approximately 15% to 20% between 1986 and 1987. This rate reduction trend is expected to continue at a slower pace for the next few months, but a catastrophic satellite failure could cause a significant increase in insurance costs in a relatively short period of time.

15.5 OTHER INSURABLE RISKS

Terrestrial network outages due to master hub damage or equipment failures resulting from an insured peril such as fire, earthquake, or hurricane may be covered under an in-place property policy of an all-risks insurance program. Space segment failures are usually excluded from standard policies. Transponder failure insurance policies cover outages in the space segment and can be purchased by microterminal (VSAT) users and vendors depending upon risk and exposures.

15.6 SAMPLE QUESTIONS TO ASK

In the event of a network outage these are some questions to ask:

Which satellite and transponder will be used to restore the network? (Caution: this may change with time as the alternate transponders may be leased or sold to others.)

Who is responsible for repointing the earth stations if required?

What are the costs and time involved for repointing?

Is there a terrestrial backup?

Will costs be incurred for alternate transponder capacity? If so, has a price been negotiated?

Are any satellite relocations planned?

If the satellite is moved, how long will it take to get to the new location?

What is the resulting financial impact of an outage?

15.7 ACKNOWLEDGMENT

Much of the material for this chapter was originally prepared by Ann M. Deering of the Space Systems Group of Johnson & Higgins, 125 Broad Street, New York, NY 10004.

GLOSSARY

ASCII Code	American Standard Code for Information Interchange—a binary code used to exchange information between computers.
Attenuation	Reduction in signal strength due to a loss of power; also, the measure of the signal loss.
Attitude Control	Maintenance of the satellite's orientation with respect to the earth and the sun.
Azimuth	The angle between an antenna beam and true north, measured along a horizontal plane.
Backoff	The process of reducing the input and output power levels of a traveling-wave tube to obtain more linear operation.
Bandwidth	The useful frequency range of a device such as a transponder (MHz).
Beamwidth	The angular coverage of an antenna beam. Earth station beams are usually specified at the half-power (or -3 dB) point. Satellite beams are based on the area to be covered.
BER	See Bit Error Rate.
Binary Code	A code which is based on only two characters: 0 and 1.
Bit	An abbreviation of "binary digit," which represents a single character in a group, either a 1 or 0.

Bit Error Rate	The fraction of a sequence of message bits that are in error. A bit error rate (BER) of 10^{-6} means that there is an average of one error per million bits.
B-MAC	A method of transmitting and scrambling television signals where MAC (Multiplexed Analog Component) signals are time-multiplexed with a digital burst containing digitized sound, video synchronizing, authorization, and information.
Boltzmann's Constant	The constant used to obtain noise power from noise temperature. Its value is 1.38×10^{-23} J/K, or -228.6 dBW/K/Hz.
BPSK	Binary Phase Shift Keying—a type of digital modulation in which the carrier phase takes on one of two possible values.
BSS	Broadcasting Satellite Service—sometimes called direct Broadcast Satellites (DBS)—transmits signals directly to homes and community reception stations.
C-band	Frequency range of 3.7–4.2 GHz for the downlink and 5.925–6.425 GHz for the uplink for domestic FSS satellites.
CCITT	International Telegraphy and Telephony Consultative Committee—a part of the International Telecommunications Union.
CDMA	Code Division Multiple Access—a multiple-access scheme where stations use spread-spectrum modulations and orthogonal codes to avoid interfering with one another.
Channel	(1) A half-circuit. (2) A radio frequency assignment. (3) A video program signal.
C/I	Carrier to Interference ratio.
C/IM	Carrier to Intermodulation ratio.
Circuit	(1) A complete (two-way) telecommunications loop. (2) Two half-circuits.
C/kT	The ratio of the carrier to the noise temperature multiplied by Boltzmann's constant. This ratio is on a per-Hertz basis and is equivalent to C/N_0.
C/kTB	The ratio of the carrier to the total noise contained in bandwidth B Hertz.

C/N	Carrier-to-Noise ratio (the same as C/kTB).
C/N_0	The ratio of the carrier-to-noise ratio on a 1-H basis.
Code Division Multiple Access	See CDMA.
Companding	The process of compressing the amplitude range of speech signals at the transmitting end of a system and expanding these levels at the receiving end.
CONUS	Continental United States
Coverage	The area on the ground contained within the satellite antenna pattern to a specified power level below the peak power radiated. This contour is often specified at the -3 dB point.
CSSB/AM	Companded Single Sideband Amplitude Modulation.
Data Base	An organized collection of data about a subject.
dB	Decibel—a logarithmic (to the base 10) representation of a ratio.
dBi	The ratio of the gain of an actual antenna to an isotropic radiator using decibels.
DBS	Direct Broadcast Satellite, see *BSS*.
dBW	The ratio of the power to one watt expressed in decibels.
Decibel	See dB.
Delta Modulation	A process for digitally encoding and transmitting an analog waveform that sends only the difference between transmissions.
Digital	Any forms of telecommunications using binary (or other) representations of data or analog level. Digitized voice or video involves sampling an analog waveform (from the mouthpiece of a telephone or the image sensor of a video camera), measuring the amplitude of each sample, and converting this measurement into a digital bitstream for transmission.
Downlink	The retransmission of a signal from a satellite transponder down to earth for reception by earth stations.
E_b/N_0	The ratio of the energy per bit to the noise density (often expressed in decibels).

EIRP	Effective Isotropically Radiated Power—equal to antenna gain multiplied by transmitted power.
Elevation	The angle between an antenna beam and the horizontal plane.
Encryption	Systematic modification of a signal to prevent unauthorized use.
Error Detection	The detection of bits in a digital data stream that have been corrupted during transmission.
FDMA	Frequency Division Multiple Access—a technique where signals from several earth stations share a satellite or transponder by using different frequencies.
FEC	Forward Error Control—adds unique codes to the digital signal at the source so errors can be detected and corrected at the receiver.
Figure of Merit (G/T)	The ratio of the gain of a receiving antenna to its noise temperature.
FM	Frequency Modulation.
Footprint	The area of coverage of a satellite receive or transmit antenna.
Forward Error Correction	See FEC.
FSS	Fixed Satellite Service—a term describing satellite communication with fixed earth station.
Gain	The inverse of attenuation. The gain, expressed in decibels, is the amplification of the signal by a device.
Geostationary	A satellite that has a circular orbit which lies in the plane of the equator, and moves about the earth's polar axis in the same direction and with the same orbital period as the earth (about 24 hours). To an observer on the ground, the satellite appears to be fixed in the sky.
Geosynchronous	A satellite that has a circular orbit which does not necessarily lie in the plane of the equator, but has the same orbital period of the earth (one sidereal day).
GHz	Gigahertz (1000 MHz).
G/T	Antenna gain to noise temperature ratio used to characterize earth stations.
Hub	The master station through which all communi-

cations to/from and between microterminals must flow.

IM — Intermodulation distortion—occurs when two (or more) signals are passed through a nonlinear device (such as an amplifier).

Inclination — The angle between the earth's equatorial plane and the orbit place of the satellite (deg.).

Intelsat — The International Telecommunications Satellite Organization.

Interface — Defines the physical and electrical link between two devices.

Interference — The unwanted interaction between a desired and one or more undesired radio frequency signals. This interference may occur between cofrequency satellite or with sharing terrestrial services.

ISDN — Integrated Services Digital Network.

ISO — International Standards Organization.

Isotropic Antenna — A theoretical antenna that radiates (or receives) the same amount of signal in all directions.

K_a-band — Frequency range of 17.7–21.2 GHz for the uplink and 27.5–30.0 GHz for the downlink for FSS satellites (''30/20 GHz'').

K_u-band — Frequency range of 11.7–12.2 GHz for the downlink and 14.0–14.5 GHz for the uplink for domestic FSS satellites (''14/12'').

LAN — Local Area Network.

LEC — Local Exchange Carrier.

LNA — Low Noise Amplifier—a preamplifier used to strengthen a weak satellite signal. The LNA is used at the feedhorn of the TVRO satellite antenna and typically features a noise temperature of 250K at 12 GHz.

LNC — Low Noise Converter—a single antenna mounted package that combines the LNA and downconverter to an intermediate frequency (e.g., 70 MHz).

Location — The satellite's assigned longitudinal position over the equator. (East or West longitude.)

LPC — Linear Predictive Coding.

Margin	The difference between a threshold and the value of the signal. If the signal level is below a receiver's threshold, the signal-to-noise ratio changes much more rapidly than the corresponding carrier-to-noise ratio. Operation at or below threshold is not advised.
Mb/s	Million of bits per second.
MHz	Millions of cycles (Hertz) per second.
Microterminal	Small earth station antennas, generally less than 2.8 m (also called VSATs).
Modem	A communications device that modulates digital signals at the transmitting end and demodulates them at the receiving end. The name is formed from the words modulator and demodulator.
Modulation	The process by which some characteristics of a high-frequency carrier signal, such as frequency, phase, or amplitude, are varied by a low-frequency informational signal.
MSS	Mobile Satellite Service—a term describing satellite communications with mobile earth stations.
Multiple Access	The ability of more than one user to use a transponder. Transponders have three basic resources: Frequency, time, and space. The frequency domain is used in FDMA. Time domain multiple access is used in TDMA by time-sharing the transponder. Space domain multiple access (SDMA) makes use of either the polarization discrimination, orthogonal digital codes (which are transparent to one another in CDMA), or through spread-spectrum (SSMA) techniques.
Noise	Extraneous signals generated in amplifiers or attenuators. Noise may also come from black body radiation of hot objects such as the sun, the moon, or the earth. The noise temperature as seen by an antenna looking into deep space is 4 K. When looking at a very dense rain cell, the perceived temperature will increase. The noise temperature of the sun caries through the solar cycle and with frequency.
Orbit Spacing	The angular separation (measured in degrees on longitude) between satellites using the same frequency and covering overlapping area (deg.).

Orthogonal Codes	Noninterfering digital codes (used in CDMA).
Orthogonal Polarization	Signals transmitted at right angles to one another.
Packet Switching	A transmission method that involves digital telecommunications from many users bundled together. Incoming digital messages are broken into short bit streams (packets) to which forward error correction bits and addresses may be added. The packet is transmitted to the destination by a variety of routes (including satellite). If one route fails the packet is rerouted.
Protocol	A set of rules and procedures for establishing and controlling conversations on a line.
PTT	Postal, Telephone, and Telegraph authority (usually government run).
Pulse Code Modulation	PCM—a time division modulation technique in which analog signals are sampled and quantized at periodic intervals. The values observed are typically represented by a coded arrangement of 8 bits of which one may be for parity.
QPSK	Quadrature Phase Shift Keying.
Rain Losses	The attenuation of a signal due to rainfall. It should be noted that the noise temperature perceived by the receiving antenna may also increase due to rain being present in the link.
RBOC	Regional Bell Operating Company.
RS-232	Specifications published by the Electronic Industries Association (EIA) for mechanical and electrical interface standards to digital communications equipment.
Saturation	The highly nonlinear operation of an amplifier in which an increase in the input produces no increase in the output.
SCPC	Single Channel Per Carrier. An FDM technique in which each channel is modulated on its own carrier for transmission.
SCPT	Single Carrier Per Transponder.
Scramble	Deliberate distortions of information of permit only authorized reception.
SDLC	Synchronous data link control. A bit-oriented protocol used as the link access procedure for ISM's SNA.

Signal-to-noise-ratio (S/N)	The power ratio of the desired signal to the noise accumulated in the transmission system from its original source through to the final point of measurement.
SNA	Systems Network Architecture.
Solar Outage	If an antenna is pointed at or near the sun, the high radiated noise level of the sun may be many times stronger than the desired signal. Solar outages occur when an antenna is looking at a satellite, and the sun passes behind or near the satellite and within the field of view of the antenna. This field of view is usually wider then the beamwidth. Solar outages occur twice a year and are predictable as to the timing for each individual site.
Spot Beam	A focused satellite signal that covers only a small region. Outside that area, the signal may be unusable.
Spread Spectrum (SSMA)	The transmission of a signal using a much wider bandwidth than would normally be required. Spread spectrum also involves the use of narrower signals that are frequently hopped through various parts of the transponder. Both techniques produce low levels of interference between the users. They also provide security in that the signals appear as though they were random noise to unauthorized earth stations. Both military and civil applications have developed.
SSPA	Solid-State Power Amplifier
Stationkeeping	The process of maintaining a satellite at a particular assigned location in the geostationary orbit.
Sun Outage	See Solar Outage.
T1 Carrier	A TDM system for carrying digital voice and data at approximately 1.544 Mb/s.
TDM	Time Division Multiplexing—a technique where several digitized signals from the same earth station share a transponder by using it at different times in the bitstream.
TDMA	Multiple access may also be achieved in a transponder by use of the time domain. In many applications the entire bandwidth is time-divided among many users. Each user transmits a short burst of information to the satellite. It receives

	bursts from distant stations and selects those time periods that contain its station identification code.
Transponder	Basically a receiver followed by a transmitter. It receives an uplink signal at one frequency, converts its frequency, amplifiers it, and retransmits it to the ground.
TT&C	Tracking, telemetry, and control (or command).
TWT	Traveling-Wave Tube.
TWTA	Traveling-Wave Tube Amplifier—an amplifier using a TWT and a power supply.
Uplink	The transmission of signals from an earth station to the satellite.
VSAT	Very Small Aperture Terminal (see Microterminal).
X.25	A set of packet-switching standards published by the CCITT.

INDEX